FUTURE
RISING
FUTURE
RISING
FUTURE
RISING
FUTURE
RISING

FUTURE
RISING
FUTURE
RISING

A Journey from the Past to the Edge of Tomorrow

ANDREW MAYNARD, PhD

Coral Gables

For permission requests, please contact the publisher at:
Mango Publishing Group
2850 S Douglas Road, 2nd Floor
Coral Gables, FL 33134 USA
info@mango.bz

For special orders, quantity sales, course adoptions and corporate sales, please email the publisher at sales@mango.bz. For trade and wholesale sales, please contact Ingram Publisher Services at customer.service@ingramcontent.com or +1.800.509.4887.

Future Rising: A Journey from the Past to the Edge of Tomorrow

LCCN: 2020933893

ISBN: 978-1-64250-263-3

BISAC: SOC037000—SOCIAL SCIENCE / Future Studies

Printed in the United States of America

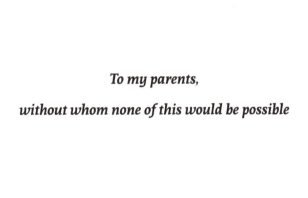

To my parents,

without whom none of this would be possible

TABLE OF CONTENTS

PART 2

UNIQUELY HUMAN

FOREWORD

My world collided with Andrew's when we were asked to join forces to host a podcast for the Interplanetary Initiative at Arizona State University. Both of us seem to share an excitement about the possibilities that exist at a place like ASU, where reimagining the future and making education accessible are the first steps to engaging students and helping them find their paths in the world.

While recording interviews for the podcast, I was struck by how illuminating his questions were, bringing perspectives from science fiction to physics to philosophy. Those perspectives permeate *Future Rising*.

The structure of this book is one of its delights. Through these short but always engaging chapters, Andrew creates a path for us to follow—and it is a trip worth taking: from light to movement, from imagination to curiosity, from possibility to hope. Each chapter reveals a new way to think about the possibilities in the world around us.

Andrew opens his book with the famous Earthrise photo, taken in 1968, an image that had a profound impact on my generation. As an astronaut who has flown in space twice on the shuttle and lived on the International Space Station, I was privileged to see our Earth from space every day for almost six months. That perspective is startling.

The Earth from space does not look fragile—the rock itself will survive long after we do. But there is a sense of our vulnerability, made palpable on my mission when we viewed the aftermath of the 2011 tsunami in Japan, only days after working with Japanese colleagues on the ground to berth their supply ship to the space station. It was comforting for us as a crew to be able to gather data about the flooding to send back to Earth. We also folded origami white cranes and sent videos of them floating in space, which played on billboards during the rebuilding to represent hope for the Japanese people.

From that vantage point, it is also clear that all of us come from one place. One planet. One home. We are the crew of Spaceship Earth and it falls on us to find a way to continue to thrive on our precious and beautiful planet. From space, it is clear that there are no borders, and nothing to stop us from being the best of collaborators, to form the best teams, to collectively bring our imaginations to bear to solve the critical problems of our times.

Imagination, as it should, underpins everything in *Future Rising*. Coupled with a sense of hope and possibility, it's what took us to space in the first place, and it is a key tool in solving the problems we face. Every section in his book gives us a different lens through which to see ourselves and others, and to explore the past, the present and what the future might hold.

To paraphrase Andrew's introduction, reading *Future Rising* really does help us "hit the metaphorical reset button to think afresh about the future." The most resounding feeling I'm left with after reading this book is hope, because in every section, Andrew shines a light and shows a path toward a future that could be ours.

CADY COLEMAN, retired NASA Astronaut

INTRODUCTION

We live in a world in turmoil. As I write, we are grappling with a future-changing global pandemic, refugees are being held in less-than-human conditions as they strive to build a better future, a growing wave of populism and nationalism is sparking another type of global epidemic—this one of mean-spirited inhumanity, and people the world over are being denied the futures they aspire to because of the narrow-mindedness of others. And over everything, there's the looming disaster of climate change, as we sacrifice long-term sustainability for short-term gain.

It's a picture of the present that doesn't bode well for the future. Yet, dire as the outlook seems, it is not inevitable. Despite appearances, our collective ability to imagine and build the future we aspire to has never been greater. With advances in science and technology, we are on the cusp of mastering biology, of creating machines that think, and of conquering seemingly-incurable diseases. Our understanding of people and society is revealing pathways toward more equitable and just futures. And we're beginning to figure out ever-more-effective ways of living sustainably within the constraints of the planet we inhabit.

And yet, unless we better understand our relationship with the future and our responsibility to it, there will remain a gaping chasm between what we are capable of and what we achieve. As a species, we are profoundly talented architects of our own future. But to do the job well, we need to get a much better grip on where we're heading, and how to ensure it's a better place than the one we came from. This is a responsibility we all face. And as we do, we each need to find our own personal threads that tie together past and future, and learn how to weave them together into a bigger picture of what humanity could become.

This, though, is no mean task. Every day, we're bombarded with information and advice in a cascade of news, commentary, and opinion that conspires to hide and tangle these threads—often leaving us feeling dazed and confused. And sometimes, we simply need to take time out—to find a still, quiet place, where we can begin to piece together a picture of the future that makes sense.

But finding such a quiet place isn't easy. We're obsessed with the future, to such an extent that it can threaten to overwhelm us. We avidly suck up news feeds and forecasts to get a glimpse of what's coming down the pike. We immerse ourselves in social media to keep up with what's on trend, and what's not. We consume book after book about the future that tells us what we're doing to destroy it, how technology will transform it, how our wrong-headedness is hampering it, and how our intelligence is manufacturing it. And we surround ourselves with science fiction stories that guide and color our ideas of how the future *might* pan out.

And yet, despite all of this, we rarely take the time to hit the reset button and think afresh about what the future is, where it came from, why it's so integral to our lives, and what our responsibilities to it are.

It's this metaphorical reset button that I set out to hit in this series of short reflections. My aim was to make them long enough to help carve out a quiet space for reflection, but short enough that they don't add to the noise. Through them, I hope to take you on a journey into what the future is, why it matters, and how we can collectively navigate toward one we aspire to as we live our lives together.

What emerges is a journey into our understanding of the future that spans history, starting at the very beginning of all things, with the celestial "big bang," and ending with our responsibility to ensure that coming generations inherit a better future than the one passed on to us. It's a journey of twists and turns, of unexpected insights, and serendipitous delights. And it's one that, step by step, builds a picture, not only of what the future is, but what our roles and responsibilities are in crafting and creating it.

Like all pictures, of course, this is one that tells a story that is far from complete. Look closely, and there are broad brushstrokes where you might expect fine detail, and curious omissions where you might expect deep insights. Yet step back, and a unique portrait of the future emerges that sheds new light on its relationship to us, and our responsibility to it.

It's also a portrait that, despite the turmoil we see around us in the present, gives us hope for a future that could far exceed our expectations—as long as we all play our part in building it.

PART 1

A JOURNEY INTO THE PAST

"No sensible decision can be made any longer without taking into account not only the world as it is, but the world as it will be."

—Isaac Asimov

1

EARTHRISE

On December 24, 1968, William Anders took one of the most influential photographs of the past hundred years. The picture was taken from the Apollo 8 mission as it orbited the moon, and it showed a startling image of the Earth, suspended in space above a bleak lunar landscape.

Anders's *Earthrise* photo captured a growing sense of our planet as a precious object that, despite its stunning beauty, was in danger of being systematically destroyed by short-sighted consumerism and greed. And it galvanized a generation to protect and preserve a future that was increasingly seen as being under threat.

In the intervening decades since Anders's photograph was first published, our ability to exploit the Earth and, with it, our future, has only escalated. As the Earth's population climbs toward eight billion, with many people still facing poverty, hunger, and substandard living conditions, we are polluting the planet and stripping it of its resources faster than ever. Climate change that's being driven by human activity is now one of the greatest threats we face. At the same time, we are beginning to overstep other environmental boundaries that help keep the world we live in on an even keel, from decreasing biodiversity to increasing ocean acidification. And this seeming disdain for the future is only being exacerbated by a growing fascination with powerful and potentially destructive new technologies.

Yet within this seemingly dystopian vision of the future, there is room for hope. Since Anders took his photograph in 1968, we've seen profound advances in potentially beneficial science and technology. As a result, more of us are living longer, healthier lives than ever before. We can now treat and manage diseases that were once considered unmanageable. The digital revolution and the internet have put more information, knowledge, and power into the hands of more people than at any previous point in history. Emerging technologies from gene editing to artificial intelligence are opening the door to potential futures that were once little more than science fiction. And threading through all of these advances, social norms and expectations are evolving around how we should behave, and what our responsibilities to future generations are.

Collectively and individually, we have more control now over how our future unfolds than ever before. But our ability to envision and engineer the future comes with almost unimaginable levels of responsibility as, together, we grapple with what we want it to look like, and how to avoid costly and potentially catastrophic mistakes.

Anders's 1968 photograph set us on a path toward powerful new ways of imagining our future. As well as showing us the majesty of our planetary home from space, he encouraged us to think of the future as an "object," something real and precious that we can and should aspire to, and to imagine the possibility of a vibrant home we would gladly bequeath to generations to come. It's a metaphor for what lies ahead that has the power to tear our eyes from the present and focus them on a future that we have a hand in designing and creating. But for it to truly transform how we think about the future, we need to go on a journey that starts billions of years in the past, when the distinction between what was and what is to come was little more than a spark in the cosmos's eye.

2

ORIGINS

Around 13.8 billion years ago, the universe as we know it came into existence. Before this, there was no future, no present, and no past, just a pinprick of potential that existed outside what we now experience as space and time.

Prior to what we think of as the "big bang," time did not exist. There was no sense of what had just happened, or what might happen next. What we think of as the future was only made possible as our present universe popped into existence, much as the cascading, irresistible flow of a flooding lake is unleashed when a dam collapses.

Because of the way our minds work, it's almost impossible for us to wrap our heads around that point of compressed space-time that existed just before the universe came into existence. And yet, over the past hundred years, scientists have, quite remarkably, been able to reconstruct those earliest moments. As a result, we now know that what we think of as "past" and "future" are merely the byproducts of the laws of physics that emerged after the big bang took place—those laws that we often take for granted, yet which govern everything from the behavior of subatomic particles to the movements of the largest galaxies.

It's sobering to think that, had some unfathomable cosmic tick not triggered the big bang, we wouldn't have a future. And yet, because of an event that scientists are still trying to understand, 13.8 billion years ago, time and space came into being and the future was born. The cosmos belched, and the scene was set for creatures like us to emerge—creatures so trapped in the resulting flow of space-time between past and future, that we can't imagine life any way.

Of course, it's easy to get mystical at this point and start to imagine what it would be like to exist outside of time. Maybe the defining essence of humans is a soul that, somehow, escaped the physics of the big bang. Or perhaps our lives are guided by celestial beings that transcend the limitations of everyday reality.

Sadly, intriguing as these ideas are, everything we currently know about the universe indicates that there is no plane of existence that transcends space and time. Every ounce of our existence, it seems, is trapped in this irresistible flow from past to future. With that cataclysmic event that started everything, the die was cast for a universe where, in practical terms, everything has a past and a future that are irrevocably tied together by the present.

It is, of course, disappointing that we can't just step in and out of time at will. But the more we understand this one-way river of intertwined space and time that we're being swept along by, the more we can learn how to nudge the future in a direction we'd like to go in, rather than the one that fate hands us. That is, if we can see where we're going.

3

LIGHT

Light, in all its forms, is such a basic part of life for most of us that it's sometimes easy to take it for granted. Light from the sun is what powers the Earth, from its geological evolution over vast timescales to its weather, the energy we rely on, and ultimately, life itself. Light has physically and metaphorically guided our planetary and evolutionary history for billions of years. And it continues to illuminate our future.

Walk into a darkened room and flick the light switch and, in most cases, what was invisible immediately becomes visible. And with it, different possible futures are revealed. These may be as mundane as avoiding stubbing a toe on a protruding piece of furniture, or as profound as—quite literally—being *enlightened* by what you can now see.

Light reveals a pathway between where we are now and where we're heading. It enables us to develop new knowledge as it illuminates the world around us. It allows us to explore how the past influences the future by observing the relationships between cause and effect. It even provides the illumination for many scholars to capture their ideas in writing, and for their students to read and benefit from these—even if that illumination is sometimes the light of a computer screen. More than this, though, light infuses our thinking about what is coming next, as we talk about "seeing" into the future, or "envisioning" it.

Yet, even before humans were on the map in the cosmic scheme of things, light was playing the role of arbiter between past and future.

As the initial maelstrom of the big bang settled down into something approaching the normality we're now familiar with, the cosmos was flooded with the fundamental particles that act as the building blocks that make up the universe and the glue that keeps it together. We're perhaps most familiar with those particles that represent visible light—the photons that are emitted from fires, light bulbs, computer screens, and, of course, the sun. But these represent just a small slice of the spectrum that scientists think of as "light." This spectrum extends all the way from intense, destructive gamma rays to long, lazy radio waves, with visible light sandwiched into a narrow band somewhere in between.

All of these forms of light form connections between the past and the future. This is perhaps most famously seen in Einstein's theory of relativity, which depends on the speed of light in a vacuum remaining the same, wherever you are and whatever you're doing.

Because light travels at a finite speed, we're still, quite remarkably, receiving signals from the very earliest moments of the universe. Incredibly, we can actually detect the afterglow of the big bang in the form of cosmic microwaves that have taken nearly fourteen billion years to reach us. These signals from the universe's past are deeply revealing of where we come from on a cosmic scale, and they help us better understand where we're ultimately heading.

But there's an aspect of light that's even more fundamental to our understanding of the future.

Light is emitted when charged particles oscillate back and forth. This is how transmitters emit radio waves. It's also why atoms emit light as the negatively charged electrons in them move between energy states.

This connection between the electrons in atoms and light turns out to be deeply relevant to the passage of time between past and future. For every oscillation, every turn of the atom-electron spinning top, emitted light waves slice like a metaphorical knife between what has just been and what's to come. Without light, there is no past and no future. And without past and future, there is no light.

Fittingly, we actually measure time using the frequency of light emitted by oscillating electrons. A single second is defined as the time it takes for nine billion, one hundred and ninety-two million, six hundred and thirty-one thousand and seven hundred and seventy oscillations of an electron transitioning between two energy orbits in a cesium atom. It's a frequency that is, sadly, too slow to be seen as visible light. But it can be picked up by a high-frequency radio receiver. And thus, light becomes the metronome that keeps time as the past transitions to the future.

But as light ticks the seconds away, it reveals yet another important aspect of the transition from past to future: movement.

4

~~~~~~

MOVEMENT

Knock a glass off a tabletop, and there's a moment when time slows. You see the glass tumbling slowly toward the hard floor, and you know that in a split second it will be transformed into a thousand shards. Or not, if your reflexes are fast enough to intervene between the future you see unfolding and the one you imagine could play out.

When it comes to the future, movement is paramount. We're constantly talking about the future in terms of movement. We "move toward" the future, we take "steps" to change it, or avoid what we imagine it might contain. We even "speed" toward the future, albeit sometimes unwillingly.

These metaphors for how we think about what's to come reflect the reality that movement is part of the natural transition from past to future.

This connection between movement and future is clearly illustrated by the falling glass. It represents a transition between what was and what, in a few seconds, is going to be. Yet the connection goes far beyond this.

As we saw previously, light results from the movement of charged particles. But movement is even more deeply

embedded in the physics of the universe. In the esoteric
world of particle physics, the building blocks of everything
around us—protons, neutrons, electrons, quarks, and
more—are held together by other particles that shuttle
back and forth between them. These form the cosmic glue
that holds everything together. It's a game of catch at an
infinitesimally small scale, where stuff sticks together
because of the constant movement of elementary particles
between the basic building blocks of the universe. Stop the
motion, and everything falls apart.

The same applies to the largest objects in the cosmos. The
Earth is in motion around the sun, which is in motion
within the galaxy, which in turn is wending its own way
between a myriad other galaxies. All are tied together by
a gravitational pull that scientists increasingly suspect is
governed by gravitational particles—gravitons—shuttling
back and forth between celestial bodies.

All of this may sound far removed from a glass falling
from a table. But without this continuous and irresistible
movement from past to future, neither the glass, the knock,
the fall, or its ultimate future, would be possible.

Movement—whether the oscillation of an electron, the
passing of a light beam, or the falling of a glass—marks the
transition between past and present, and points toward
the future. Yet movement can only occur because of that

intangible yet ever-present essence that envelops us all: time.

5

TIME

In 1988, the late British scientist Stephen Hawking published what is possibly the most successful book never to be read. Although *A Brief History of Time* sold like hotcakes, it's tough going for most readers. Yet the book tapped into our near-insatiable fascination with time, and how it both constrains our lives and opens up new possibilities.

Time dominates our lives. With very few exceptions, we are deeply aware of its passing, and the ways that it guides and molds us. The Earth marks out time in its orbit around the sun with every passing day and year. We wake, work, and sleep to a given rhythm of time. We're obsessed with what we did with our time in the past, and what we're going to do with it in the future. We celebrate birth as a new chapter in our time-driven lives. We worry about death as the end of a chapter, along with what—if anything—comes next. And we surround ourselves with devices that remind us of the inexorable passing of time, from our watches and clocks, to our phones, laptops, fitness trackers, and every conceivable manner of internet-connected device.

We are, at every level, creatures of time, immersed in it, obsessed by it, yet unable to control it.

This, not surprisingly, deeply colors our visions of the future. How we experience time allows us to imagine what the future *might* be like. But it also throws up an opaque veil between us and the future. It offers us tantalizing glimpses of what might occur in the future, while holding us back from experiencing it until we get there.

Unlike the many time-related fantasies explored in science fiction, we sadly cannot move back and forth in time, beyond the second-by-second movement forward to which everyone is subject.

Of course, there's an exception to every rule, and the exception here is that, at near-light speeds and in the presence of massive gravitational fields (a black hole for instance), time no longer ticks along at the same rate. But for the vast majority of us, we're stuck on a one-way trip into the future, with all of us on the same track.

And yet, the same science that prevents us from flitting back and forth through time occasionally allows us to see—albeit darkly—what might potentially occur in the future. This has nothing to do with us mystically overcoming the shackles of time. Rather, it's based on how the laws of physics allow us to predict what's likely to happen, based on what's occurred in the past.

And one of the more intriguing of these predictions
involves how the passage of time is causing the universe
and everything in it to slowly run down, through the
accumulation of "entropy."

6

ENTROPY

In a 2018 ranking of US states by road quality, Florida and Hawaii did pretty well, but my former state of Michigan did not. In fact, the report confirmed Michigan's dubious status as the pothole capital of the country—something my bones can vigorously attest to!

Potholes may seem a far cry from the future of the universe, yet they are both tied together by a natural tendency that threads through the cosmos: entropy.

Entropy is one of those concepts that people often invoke when they try to explain life, the universe, and everything, yet is rarely understood—much to the chagrin of physicists the world over. Despite this, it plays an important role in determining how and why the universe behaves as it does, as it slips from past to future.

Entropy is related to the amount of usable energy in a system—the energy that you can actually put to work to achieve something. The greater the entropy, the less usable energy there is. The idea goes that, if you have an imbalance in energy between one object and another—say, for instance, between a cool Earth orbiting a searing hot sun, or a really hot cup of coffee and a pair of cold hands—you can make that energy work for you by moving it from one place to another. But when the imbalance has been eliminated, you're done. You can no more make use of energy when any energy difference has gone than you can make water flow uphill. As energy is used, imbalances are reduced, and entropy increases. And potholes, it turns out, are a symptom of this.

Potholes are, unfortunately, part of the natural future state of roads. They're what roads aspire to, if only people would leave them alone. And they are an inevitable outcome of entropy. To make sense of this, think of roads as objects that represent massive amounts of stored energy in the form of the work that goes into making them straight, smooth, and durable. As a result, there's an energy imbalance between them, the environment, and the tires that they're constantly being pounded by. As that energy difference is evened out, with a little help from the elements and daily wear and tear, cracks and crevices form and ultimately turn into potholes. And as they do, entropy increases.

Without regular repairs, our roads would simply get more and more potholed until they become impossible to use. It's a very real and frustratingly tangible example of entropy as it acts as the inexorable "arrow of time." And it's one that reminds us that, not only is the future inevitable, but when left to its own devices, it's likely to be rougher than the past.

This is the same process that many scientists believe is leading to the universe winding down. Experts are still uncertain whether we're ultimately heading for a big crunch—a fitting mirror to the big bang, where everything contracts back into a pre-big-bang singularity—or a big freeze, where the universe simply slows down and stops. Many scientists suspect that increasing entropy will ultimately lead to what they refer to as "heat death," where all the usable energy in the universe has been, not to put too fine a point on it, used up.

It's the celestial equivalent of the universe becoming so potholed that it's impossible to drive on it any more.

In the grand scheme of things, this is one future that the laws of physics indicate is likely to occur—a predictable pathway toward a dead universe, where increasing entropy has sucked the marrow out of every atom and every molecule. It's a default future that has little time for the complexity of living organisms. Yet, far from slipping toward heat death, Planet Earth has somehow spawned organisms that can seemingly reverse the universe-wide flow of entropy. And because of this, we have emerged as creatures that can not only imagine the future, but also intentionally alter it.

7

EMERGENCE

In 2014, the Massachusetts Institute of Technology physicist Jeremy England shook the scientific world with an audacious new explanation for the emergence of life on Earth. And in doing so, he opened the door to an intriguing explanation for how we might have gotten here, and why we're so obsessed with understanding and changing the future.

One of the side effects of entropy is that things generally get more disorganized as the universe ages and usable energy decreases. It's the celestial equivalent of a kitchen that becomes increasingly messy if no one's putting in the effort to clean it, or the degradation of roads if road crews aren't constantly fixing them.

According to the laws of physics, the universe's future should be getting increasingly chaotic. And yet, life seems to fly in the face of this conventional wisdom. Compared to the primordial soup that existed billions of years ago, humans are an exquisite manifestation of organization and complexity. Our bodies and minds are fantastically intricate biological entities that are not only incredibly complex, but are also able to create order out of chaos. We are, in effect, localized anti-entropy machines that have the ability to change the future from its default mode to something entirely different.

Life's ability to buck this celestial arrow of time has puzzled scientists for decades. There are, as you would expect, plenty of explanations that scientists weave around this seeming anomaly. For instance, most scientists would argue that, while net entropy always increases, there's nothing to stop temporary decreases at a more local level—such as the emergence of living organisms on a favorable planet. And yet, life as we know it still seems to lie so far from the apparent comfort zone of the universe that it sometimes feels like our explanations of how we came to be here are, at best, long shots.

In contrast, England came up with a possible explanation for this seeming anomaly which suggests that living organisms could be a feature of the universe we live in, rather than an exception. And, if right, his ideas could fundamentally alter our understanding of life, intelligence, and the future.

England argues that the universe is "programmed" to reach its ultimate future of heat death as fast as possible—that point where increasing entropy has eliminated all usable energy. Under normal circumstances, the speed with which this journey occurs would be limited by relatively conventional physics. But what if there were shortcuts that could accelerate this celestial decline, and get the universe to where it's going even faster?

According to England, the emergence of life may well be one of these shortcuts.

Living organisms have an amazing ability to convert energy into less usable forms, as they suck in high-energy resources, and leave behind lower-energy "excretions" (usually in the form of heat). In this way, plants and animals are entropy-accelerating machines as they burn through the energy they receive from the sun and their surrounding environment. And humans, with their creativity and inventiveness, take this to a whole new level.

To get a sense of the scale of the entropy-acceleration we're capable of, you only need to look at how a global population approaching eight billion is stripping the world of its usable energy resources, and leaving a trail of chaos and destruction behind it. And when you factor in our ability to invent ever more powerful and rapid ways of creating chaos, from guns and explosives to hydrogen bombs, you have to begrudgingly admire the universe for coming up with such an ingenious shortcut for accelerating the rate at which entropy increases.

England's ideas are controversial and as yet unproven. Yet they are compelling. They grapple with the physics of systems that are a long way from thermal (or energy) equilibrium. And in doing so, they are hinting at how the laws of the universe might throw up pockets of order that seem to defy the flow of entropy, because they ultimately accelerate the journey toward its inevitable future. They also begin to create a framework that may help us better understand the emergence of life and human intelligence, and even our tendency to cause chaos in the name of progress.

If England is right, we may all be part of a celestial shortcut that is nudging the universe ever faster toward its ultimate fate. It's a shortcut that began, however, long before humans appeared on the scene, with the earliest emergence of living organisms.

8

EVOLUTION

Four billion years ago, the Earth was a barren planet circling an undistinguished sun in a backwater of an unremarkable galaxy. Then something changed.

Whatever sparked that initial series of events, a hole was punched in the universe's slow, measured amble toward heat death. As organic molecules formed and coalesced, and the first living organisms began to appear, there was no conscious sense of past and future. And yet, the stage was set for a remarkable journey, as life was driven along toward a future that was always slightly better than what had gone before.

We now know that a large part of this driving force was natural selection. Those earliest organisms inherited a unique adaptability that was encoded within a quite remarkable molecule—the DNA that is part and parcel of all living creatures. Across generations, those organisms that were able to survive and adapt in a harsh and changing environment passed the genetic secrets of their success onto their offspring. And as circumstances changed, so did their DNA, whether through natural mutations, interbreeding, or other ways of exchanging and coopting genetic advantages.

DNA turned out to be an incredibly powerful entropy accelerator. Fed by heat, chemical energy, and ionizing radiation, it became the defining base code of increasingly advanced organisms that were ever more adept at making use of the energy around them and discarding it in a slightly less-usable form.

Through natural selection and genetic mutation, DNA began to encapsulate the blueprints of organisms that were increasingly complex and sophisticated. It's a process that is startlingly elegant in its simplicity: randomly tweak the code, naturally select the organisms with the traits that increase their chances of survival, multiply, and repeat.

In this way, evolution appears to reverse the universal flow of entropy locally, as organisms and the environments they inhabit become increasingly ordered and complex. But in the grand scheme of things, it accelerates the conversion of energy from usable forms to unusable ones, and in doing so, it speeds up the rate at which universal entropy increases.

If things were different, this might have been the end of the story. The world could have been filled with constantly evolving plants and animals that were finely tuned to accelerate the rate of increase in entropy, but were no more aware of their future than the microbes in your gut.

Yet the universe had another trick up its sleeve, and this was the emergence of creatures that had the unique ability to anticipate the future, and to adjust their behavior accordingly.

9

ANTICIPATION

In 2010, Paul the octopus shot to fame for his uncanny ability to seemingly predict the outcome of World Cup soccer games. A resident at Sea Life Oberhausen in western Germany, Paul correctly predicted the outcomes of all seven of the German team's games, including their loss to Spain in the semifinals.

It's hard to see Paul's success as anything other than a complete fluke. As a tank-bound cephalopod, he simply didn't have the relevant data at his tentacle-tips, or the mental acuity to process it even if he did. And yet, in a curious way, Paul's story captures just how important anticipation is to how we think about the future, and how we end up responding to what we foresee.

Anticipation of the future is something that's typically seen in higher-order organisms. It's an evolved capacity that reflects an awareness that there's a future beyond the present where potentially good or bad things can happen. And it is part of a suite of abilities that lay the foundations that enable organisms to not only envision the future, but to plan for it as well.

In Paul's case, his anticipation was, disappointingly, more to do with food than football. Before each game, he was presented with two clear boxes, each with a tasty mussel in it. Each box sported the flag of a different soccer team. Despite the beliefs of his fans, Paul was simply smart enough to anticipate the future satisfaction of eating the food he could see, and everything else was sheer luck.

This may seem trivial in the broader scheme of things, but the evolutionary progression from organisms that lived random lives, then died, to those that could anticipate the future and act accordingly, was a profound one. It changed how animals acquired their food, and it altered the behavior of those that were valued as food. Anticipation became an essential part of the evolutionary survival toolkit, as it allowed organisms to peer into the near future and work out what their best course of action was to either avoid it, or embrace it.

This future-looking evolutionary trait is seen in most higher-order animals, and even in some communities of social insects, such as bees and ants. And of course, it's a defining feature of humans. We're constantly anticipating the future and imagining how it's going to affect us. Whether we're driving, putting in a day's work, working out, or simply working out how we're going to get to the end of the week, we're anticipating the future and adapting our behavior accordingly. It's what keeps us moving forward. And it's what leads to our obsession with sports among other things, as we eagerly anticipate the tangible outcomes of clashes between opposing players.

Paul may not have actually been able to anticipate the outcomes of soccer games, but it was anticipation of the results that led his fans to suspend their belief and accept his "predictions." This anticipation, though, was also due to another biologically inherited "future-sense" that we all have, and one that leads to us sometimes favoring views of the future that aren't always grounded in reality: instinct.

10

INSTINCT

People who write and talk about risk perception for a living often tell a story that goes something like this: Someone is walking along a path, and they see a long, sinuous shape on the ground in front of them. Their instinctive response, before any rational thought or logical analysis has kicked in, is to freeze, fight, or run away...until they discover to their embarrassment that what their subconscious interpreted as a snake is, in fact, just a stick.

Eons ago, our primitive brain developed responses to avoid things that might harm us, snakes being one of them. Through the process of natural selection, animals that were predisposed to avoid dangers, such as objects that looked like snakes, lived long enough to pass on their genetic predisposition to the next generation. And so on.

As we've moved up the evolutionary ladder, we've retained many of these inbuilt responses. And thankfully so—these biological instincts provide us with a brilliant survival mechanism that enables our subconscious brain to peer into the near future and avoid potential dangers it sees there.

Our instinctive behaviors are both complex and nuanced. They're a combination of genetically programmed responses—a biological tendency to adopt certain ways of behaving—and the ability of our brains to assimilate conscious learning into unconscious behavior. Behavioral scientists refer to these as system 1 and system 2 thinking. But at its heart, instinct is our innate ability to foresee possible futures, and to act on them without conscious thought.

Much as we like to think of ourselves as rational creatures, our lives are still deeply influenced by instinct. We even take pride in "going with our gut," or "following our intuition." Both are part of our built-in "future-senses," those bits of us that make us feel that we can see beyond the present and peer around the corner of time. And thank goodness we have these, as there are times when our conscious brain does an appallingly bad job of making sense of signals that our unconscious brain is screaming at us to take note of.

And yet, when it comes to navigating the future, our instinct can be a liability. Instinct relies on the future being similar to the past, and predictable based on what's happened time and time again. But humans have put a huge wrench in this biological master plan as we've developed the ability to change the future faster than any evolutionary process can accommodate.

Over the past two hundred years, the rate of technological change has been accelerating at breakneck speed. We've invented steam power, electricity generation, mass production, synthetic chemicals, computers, the internet, social media, gene editing, artificial intelligence, space flight, even the ability to "geoengineer" the entire planet. And in doing so, we've created a world that our instincts are increasingly poorly equipped to handle.

These instincts, which have their roots in an evolutionary provenance that long predates the Enlightenment or the Industrial Revolution, make us predisposed to believe things that *feel* right, yet aren't necessarily supported by evidence. This is not to say that feelings-based decisions are wrong— far from it. But when they run counter to evidence, or re- enforce deep biases in how we use evidence, they can derail our ability to achieve what we set out to.

Impressive as they are, our gut instincts—our evolutionarily honed future-senses—are easily dazed and confused by the future we're creating. One way out of this conundrum is to kick our finely evolved conscious mind into gear as we develop a clearer understanding of how the past, present, and future are connected. And this means learning to draw lines between how future effects are linked to past causes.

11

CAUSALITY

Causality forms the bedrock of how the universe works. Putting aside the first tumultuous moments after the big bang and some of the more esoteric oddities of quantum physics, the future largely looks the way it does because of events that happened in the past. No matter how convoluted and complex the threads tying the past, present, and future become, each past action sends ripples into the future that spark a cascade of sympathetic reactions.

We learn this pretty quickly growing up. Grab a hot pan, and it will burn. Eat putrid meat, and you'll get sick. Needlessly insult people, and you'll get the cold shoulder treatment.

This line between cause and effect means the future is molded and crafted by what's happened in the past. Understanding this, we can begin to plan for what's heading our way, and even begin to entertain the idea of influencing it. That is, if we can work out which "cause" levers in the present lead to desired future "effects."

Making use of these threads between cause and effect is, of course, what underpins modern science and engineering. Scientists are remarkably good at asking "what if" questions that begin to unpack what the results of a set of actions or events might be, and using these to create theories and models that enable the future to be predicted. This is what scientific theories like Newton's laws of motion and Einstein's theory of relativity do.

But consciously understanding, measuring, modeling, and theorizing about cause and effect is only half the story. The other half is what we do with this knowledge. And this is where science and engineering enable us to use our conscious understanding of cause and effect to not only predict the future, but take a stab at changing it. It's this ability to make causality work for us that is enabling us to feed a growing population, to control and eradicate devastating diseases, and to elevate the quality of life for billions of people.

On the flip side, it's also what has enabled us to devise increasingly effective ways of harming the environment, and abusing and killing people. We should never assume that our mastery of cause and effect is, by default, benign.

Yet, while our ability to utilize causality has had a profound impact on humanity's capacity to predict and control the future, this ability depends on something even more basic: our ability to remember and recall what happened in the past.

12

MEMORY

Imagine not being able to remember what happened yesterday, an hour ago, or even a minute ago. Your life would exist in a small sliver of the present, with no understanding of where you've come from or where you're going. As a result, your sense of the future, and your agency over it, would be limited to the point of being almost nonexistent.

Unlikely as this might sound, this is the situation that British musician Clive Wearing found himself in after contracting *Herpes simplex* encephalitis in 1985. The infection led to highly unusual damage to his hippocampus, making him incapable of forming new memories.

Wearing's case is an extreme example of anterograde amnesia. While he can remember some things from his previous life, current memories only last for a matter of seconds. Because of this, he is stuck between past and future, always in the present, but never able to move on.

Wearing's story is a sobering reminder of the importance of memory in how we see the future, and how we begin to think about ways to navigate toward it, or even change it. If we have no memory of what's happened in the near past, we have no way of connecting effects we observe to what caused them. And this in turn means that we cannot begin to understand how our actions potentially influence the future.

This can be seen through a hypothetical example of learning a seemingly simple task in the absence of memory: learning for the first time how to open a door.

A door stands as both a metaphorical barrier and an opportunity along the pathway between the past and the future. Closed, it bars the way to what lies on the other side. But when opened, it reveals new possibilities.

First, though, you need to open the door.

Imagine for a moment that you have no concept of what a doorknob is or what it does, and you have no short-term memory retention. You're faced with a smooth surface with a circular protrusion sticking out of it. You may have previously been shown how this works, but you don't remember.

You might try a few things—pushing it, pulling it, sliding it, even turning it. But here's the kicker: Whatever you do, you immediately forget. And so, even if you begin to turn the knob, and the latch begins to open, the next second you'll be trying something else, oblivious to the fact that, if you could remember what you were doing, you'd have cracked the problem.

And it gets worse. Even if you do manage to open the door, you will have no recollection of how to do it again, because you have no memory of the cause that led to the effect.

It's a somewhat simple example, but it does demonstrate that, even in the smallest tasks, memory is essential to the way we think about and navigate toward the future. It's what allows us to build models in our mind of how the past and present are connected to what lies ahead of us. Memory is an amazing evolutionary gift that has given us the ability to chart a pathway toward futures we desire, while avoiding those we don't. Without it, we are helplessly caught in a present that we are powerless to control.

And yet, memory on its own takes us only so far toward navigating and designing the future. It gives us the capacity to construct visions of the future, and pathways to get there. But this capacity is redundant unless we have the ability to fill it with useful information, together with the skills necessary to make use of it.

13

LEARNING

I was in my mid-thirties when I moved with my family from England to the US. And because of this, I was blissfully unaware of the dangers of one of the country's more ubiquitous plants—poison ivy. Yet it only took one painful encounter to teach me the wisdom of avoiding it.

Our ability to learn is one of the most transformative skills evolution has bestowed on us. This ability to not only remember, but to associate an event with an outcome and to anticipate the consequences of our actions, uniquely equips us to imagine a future that is different from the present, and to work toward creating it.

Learning is what begins to carry us beyond instinct and allows us to start intentionally crafting the future. It's what enabled me to ensure a future that did not involve me blundering into poison ivy, and one where I learned how to identify and avoid the offending plant. And on a much grander scale, it's what's driven along every invention that's propelled humanity into the future over the past ten thousand years, from taming fire to the latest advances in space flight.

Learning is what ties together anticipation, causality, and memory, as we shrug off the shackles of instinct in our quest to build the future. And it encompasses an incredibly powerful and diverse set of evolved skills that, in turn, color our perception of the future.

Perhaps one of the most familiar tools in this learning skill set—and the one that taught me the evils of poison ivy—is learning by trial and error. Anyone who's watched a young child learn knows how effective the process of trying, failing, and trying again is in developing new skills. It's this form of learning that fueled the Industrial Revolution, as inventors persevered in trying different ways to harness the power of water, steam, and later, electricity. It's what enabled scientists like Michael Faraday to harness the power of electromagnetic induction. It underpinned the tenacity of Edison as he strove to invent a viable electric light bulb. And it's deeply embedded in the entrepreneur's mantra of "fail fast, fail forward."

Learning by trial and error is one of the ways we close the gap between how we *think* the world works and how it actually works, as we strive to create a future that's different from the past. It's not the only form of learning, by any stretch of the imagination. And to be effective, it needs to be accompanied by the ability to predict what might happen in the future, based on previous experience. But it is foundational to many of the ways in which we develop and apply new knowledge.

And yet, for learning in any form to be useful, it has to involve more than simply developing new knowledge. To work for us in crafting the future, the knowledge that comes from learning needs to be applied. But first, there needs to be intent, and the conscious resolve to bring about change through what we know, and how we imagine things might be different.

14

INTENTIONALITY

The question of whether the future is set in stone, or is malleable and designable, is as old as human civilization itself. And it's one that provides rich pickings for science fiction, including the 1984 sci-fi movie *The Terminator*.

The original *Terminator* movie takes place in a future where super-intelligent robots are at war with humanity, and in a bid to finally stamp out their adversary, they send a "terminator" robot back in time to eliminate the future leader of the human resistance.

As time-warping sci-fi movie plots go, it's as convoluted as it is implausible. Yet, beneath the rather fantastical storyline, the movie grapples with the conundrum of whether the future is set, or whether we somehow have the ability to alter it by what we do in the present.

Not surprisingly, the movie's overarching message—and one that is reinforced in the 2019 sequel *Terminator: Dark Fate*—is that we have the power to change the future. In the words of John Connor, who Arnold Schwarzenegger's Terminator was sent back to eliminate, "The future is not set. There is no fate but what we make for ourselves."

Yet there's a catch—both in the film and in real life—
in that we can only change the future for the better
if we consciously decide to do so. There has to be an
intentionality to our actions if we want to see a future
emerge that fits our dreams and aspirations.

Intentionality is the connective tissue between learning
and outcomes. It's the link between observing which levers
in the present can be used to nudge the future in different
directions, and having the wherewithal to actually pull
them. Intentionality is the difference between watching
people starving on news feeds, or being aware of the
scourge of cancer, and actually doing something about it.
And it depends critically on the realization that the future
is, indeed, not set.

Intentionality draws on an awareness of what the future
might look like, and leads to responses designed to nudge
the future closer to what we would like it to be. But because
intentional acts represent an investment of time, energy,
and reputation, we have to be pretty sure of ourselves before
we commit. We have to believe that the future we hope for is
possible, and that we have some ability to influence it. Few
people are willing to make sacrifices to bring about a future
that, in their mind's eye, isn't a real, concrete likelihood.

But what if our visions of the future and how to get there
are so flawed that they inadvertently lead to a place filled

with pain and suffering? Or worse, do so because of malicious intent?

Sadly, there will always be those who set out to intentionally create a future for themselves that is built on the pain and suffering of others. Yet I have to believe that, for most people, there is a willingness to try to build a future that benefits as many others as possible. The problem is, if our collective vision is misguided and our understanding of the tools we have at our disposal is misinformed, we can easily do more harm than good, no matter how intentional we are.

The challenge, then, is how we narrow the gap between what we hope for and what we end up getting. How do we better ensure that we are the masters of our collective fate? Here, evolution has given us one of our most precious gifts: our intelligence.

15

INTELLIGENCE

Imagine the scenario: A salesperson cold-calls you and persuades you to listen to their pitch. They present you with three visions of the future, and tell you that, for a small donation, you can be a part of building the one that most appeals to you.

The first future is one where everyone lives in harmony with nature, where there's no pollution, no sickness, and no unhappiness. The second is a future where the streets are quite literally paved with gold, and everyone's a millionaire. And in the third, all political differences have been put aside for the greater good, and everyone's pulling together to make the world a better place. All you have to do is pick the future of your dreams and commit $10 per month for the rest of your life, and watch your investment grow. The futures are beautiful, the price is right, so why would you *not* sign the contract?

I'd like to believe that most people would see through such a scam because they're smart enough to understand the impossibility of what's being offered, as well as the underlying intent of the salesperson to deceive them. But obvious as it seems, such insight is only possible because of our intelligence.

Intelligence is possibly one of the most important attributes evolution has imbued us with. It's what enables us to come up with creative schemes to build the future we desire, just as it helps us avoid scams designed to rob us of our future. Intelligence is what allows us to solidify in our mind a vision of the future as something that can be crafted, designed, engineered, and valued. And it's what helps us ensure that intentionality and learning carry us in the direction we want, without leading to too many unexpected surprises.

Yet, essential as our intelligence is to imagining and building the future, it's surprisingly hard to pin down precisely what we mean by it.

On one hand, it's easy to think of intelligence as a combination of memory, learning, and application that, together, enable us to solve problems. It's this type of intelligence that lies behind many of the inventions we rely on. An inventor, for instance, may remember that round objects roll, and learns that placing something on top of a round object helps move it from point A to point B. Before you know it, you have bicycles, trains, cars, and an epidemic of e-scooters.

This practical, problem-solving form of intelligence underpins much of modern science and engineering. And as something that helps us design and build new futures, it's a powerful one. But there are many other types of intelligence that are relevant to future-building. These include intelligence that leads to the creation of music and other art forms which reveal insights and possibilities that would otherwise remain hidden. Or intelligence that enables us to understand and respond to each other emotionally and socially.

Because what we think of as intelligence is so multifaceted and elusive, how we define it is still surprisingly contentious. This tension arises in part because our notions of intelligence are deeply tied to our personal visions of the future and how to get there. So, if you value a technology-enabled future, your idea of intelligence will more likely be one that's grounded in logic, rationality, and science. Or, if you are fixated on a future dominated by economic growth, you'll likely value concepts of intelligence that are rooted in translating knowledge into power and profit.

On the other hand, if you value a future that's environmentally sustainable, or one where health and happiness are more important than power and profit, you're more likely to think of intelligence as a complex mix of empathetic, artistic, and inspirational traits that help transform your aspirations into reality.

Despite these differences, intelligence emerges as that special something, the "secret sauce" if you like, that gives us the tools and ability to transform what we can imagine into the future we hope for. It's an engine of change that helps us craft and create the future. But like all engines, it needs fuel. And a vital ingredient in that fuel is the knowledge that comes from learning.

16

KNOWLEDGE

In 1962, John F. Kennedy galvanized a nation into creating a future that had previously been the stuff of science fiction, as he inspired Americans to "...go to the moon in this decade and do the other things, not because they are easy, but because they are hard." His vision was realized seven years later as, on July 20, 1969, the Apollo Lunar Module Eagle landed on the moon and Neil Armstrong took "one small step for man, one giant leap for mankind."

Kennedy, and everyone who worked on the Apollo mission, took a bet on changing the future and setting humanity on a new course. But when Apollo 11 blasted off, how did they know that there was a high chance of its occupants not only walking on the moon, but returning safely to Earth? One thing that's certain is that they didn't just say, "We don't know that it won't work, so let's just try it and see."

Memory and learning reveal to us that it's possible to change the future, while intentionality provides us with an impetus to act on this possibility. Yet if we don't know what we're doing, the chances of breaking something along the way are pretty high.

Intelligence is what helps us better understand the path between the present and the future. But it's knowledge— generated through learning—that helps us to move safely and effectively down this path. When combined with our intelligence, knowledge helps us begin to connect cause with effect, and to create the models and tools that allow us to make use of these connections. It's what makes the difference between actions that have unpredictable and possibly damaging consequences, no matter what the intent, and those that help move us toward the future we are trying to build. And, just like intelligence, it comes in many shapes and sizes.

In the case of the space program of 1960s America, technical knowledge was the currency of success. Scientists and engineers worked tirelessly to develop and test materials, devices, and systems that could safely deliver a crew to the moon and back. Lives depended on knowledge that not only connected myriad causes to a plethora of effects, but extended to events that had never previously been experienced. What emerged was a deep and powerful intertwining of intelligence and knowledge that enabled mission leaders to predict the future of a journey into the unknown with impressive accuracy. But this projected future—this possibility made real in the mind's eye of the American people—also had a profound impact on the type of knowledge that was developed.

Much of this knowledge was technical in nature, but not all. As the space program progressed, we learned about ourselves as individuals and as a society. We developed new ways of thinking about the future and our place in it. We were inspired to imagine what it would be like to travel to other planets and beyond, and not only what it might take to get there, but how we might be changed in the process.

The ways our visions of humanity's presence in space have unfolded since Kennedy's 1962 speech encapsulate many different forms of how we "know"—how we make sense of where we've come from and where we're going, and how we navigate our journey along this path. But underpinning much of this is our ability to identify and fill important gaps in our knowledge. And one of the more powerful attributes we have as a species to achieve this is our ability to reason.

17

REASON

One of the most profound aspects of being human is our ability to predict the future. Of course, there are deep limits to our capacity to see into the future. Chance and randomness tend to throw a wrench into our skill at peering into the unknown, as do the boundaries of our intelligence and knowledge. Yet, every day, we use our reason to anticipate the twists and turns life throws at us, and to adroitly navigate them.

Reason, unlike instinct or intuition, is the culmination of our ability to observe, learn, recognize gaps in our knowledge, fill them, and develop an understanding of how the past and present are connected. It gives us a window into the future that not only enables us to predict what happens next in many cases, but to prepare for what's unfolding, and even to alter it.

This capacity to think about and respond to what the future may hold is not unique to humans. There's growing evidence that a number of animals are able to reason their way through solving simple problems. Yet it's something that has become so advanced in us that it sets us apart from other species.

Our ability to reason is what helps us imagine the possible outcomes of events and actions, and to focus on the more plausible ones. It's a combination of observation, learning, and mental gymnastics, all tied together by our intelligence, that enables our brains to construct future-predicting "if-then" statements that keep us alive and kicking as we move forward.

This ability to peer into the future through the power of reason is so enmeshed in our everyday lives that it's easy to overlook how astounding it is. Its roots lie in the survival mechanisms we've evolved and inherited—a side effect of our biologically encoded instinct to stay alive. But over time, our capacity for reason has grown to the point where it now enables us to envision the future we want, and to work out, step by step, how to build it.

Reason lies at the heart of modern science, as we systematically learn how the universe works, and use this to predict what it's going to do next. It enables us to begin mapping out the consequences of our actions, even if those consequences take time to show themselves, as in the case of human-caused climate change. And it allows us to translate our wildest dreams into concrete realities, whether these entail going to Mars, creating new virtual worlds, or building a better society.

And yet, stupendously powerful as this attribute is, reason can be blind to the future. Reason is what tells us that 2 +

2 will always equal 4, and that the sun will always rise in the east and set in the west. But it can lead to us struggling when we don't have all the information necessary to predict the future, or when the threads tying the present to the future are so complex and convoluted that they defy analysis. And it most definitely runs into difficulties when all of our wonderful human idiosyncrasies are thrown into the mix, and it becomes clear that the past and future are tied together by more than reason alone.

PART 2

UNIQUELY HUMAN

"The future belongs to those who believe in the beauty of their dreams."

—Eleanor Roosevelt

18

FEELINGS

Take a moment to think of three things that bring you joy. They can be memories, or places, people, possessions— pretty much anything. But each one should be something that makes it worth waking up each day and facing the world.

The chances are that each of these items evokes a complex set of feelings and emotions. Despite our intelligence and our ability to reason, it's more often than not what we *feel* that brings value to what we do.

Of course, what we feel, and how this affects our lives, is deeply intertwined with our intellect. Yet our feelings are an aspect of being human that often seem to transcend mere facts and figures, and that profoundly affect what we consider to be of value—especially when it comes to human lives. And as a consequence, what we *feel* about the future is just as important in determining how we design and create it as what we *think* about it.

For example, consider the sort of future you would like to see twenty years from now. What would you want to be different from the present? What new capabilities and behaviors would you like to see?

Now think about how this future makes you feel: excited, enthusiastic, impatient? And how does the possibility of *not* achieving this make you feel, especially if someone actively prevents it happening: sad, angry, outraged?

Then, consider how these feelings affect what you're prepared to do to help build a future that you want to see come about, rather than one you don't.

While our intellect can help us envision different futures and imagine ways of creating and building them, it's what we feel—and what we hold to be of value—that often determines the futures we are willing to invest in. In this way, how we feel about the future we hold in our mind's eye becomes a guiding light to how we live our lives in the present.

Reason enables us to develop plausible and powerfully predictive mental models of the future. But it's our feelings that guide us toward the models we prefer, and that influence how we embellish these models and ultimately make use of them. However, our feelings aren't always reliable. They're rooted in those parts of our brain that deal more with our instinctive, emotional responses to our surroundings and our survival, rather than a rational analysis of what we can perceive. They thrive on our capacity to fill the gaps in what we don't know with stories that make sense, but aren't necessarily true. And because our feelings so often override our intellect, they can get us into trouble.

Yet the complex and convoluted fusion of reason and feelings that each and every one of us represents gives us a quite startling ability to reach out beyond what we know as we strive to create the future we desire. It's this combination that enables us to escape the chains of short-sighted rationality and imagine futures that transcend what we currently think possible. Feelings and reason together form the source of our creativity and imagination, and our ability to translate these into ideas that bring us closer to a future that is markedly different from the present.

They are also at the heart of our unerring ability to believe in a future that has no rational-seeming basis, and to have faith in what we cannot see and do not understand.

19

FAITH

If reason is deducing what lies in the future based on the logical extension of what we know of the past and present, faith is, in some ways, its antithesis. Faith depends on believing in what lies in the future without supporting evidence, or even despite the evidence in front of us. To a rationalist, faith makes little sense. And yet this ability to believe in a future for which we have little or no evidence seems to be hardwired into the human psyche, irrespective of whether we are religious or not. And as a consequence, it has a profound impact on how we envision the future, and how we live our lives in the present.

Just a few miles from where I live, there's a neighborhood with the rather grandiose name of Valhalla. It's a residential area, where realtors, without any sense of irony, advertise "Tours of Valhalla." It's also, perhaps fittingly, the place where L. Ron Hubbard founded Scientology.

The original Valhalla is, of course, the great hall within Norse mythology where heroes killed in battle are said to live it up after their death. I suspect that the Valhalla just down the road is a little tame by comparison. But both are testaments to the power of belief and faith in connecting the future to the present.

According to Norse mythology, Valhalla was a reward for courage in battle—a vision of the future that, if the stories have any truth to them, profoundly influenced behavior in the present. Warriors accepted on good faith that a glorious future awaited them after death in battle that would wipe away the pain and guilt of their real-world wounds and atrocities. There was no intellect or reason here, just the reality that persuasive people and compelling stories can hold incredible sway over our visions of the future.

On the face of it, Scientology is very different from Norse mythology. It's a belief system that claims to be based on rational thinking and scientific methodology, even though these lead to some rather unconventional ideas— including the claim that humans are the manifestation of the immortal souls of an extraterrestrial race. Yet as with Valhalla, Scientology inculcates in its followers a vision of a future that transcends death, and a faith where this believed-in future deeply affects the way they live their lives in the present.

Scientology and Norse mythology are extreme examples of faith and its relationship to the future. Yet they serve as a reminder of just how deeply faith in the unknown, and often unknowable, can color our perspective on our future. Almost every religious tradition has a vision of the future that deeply affects the way its followers live their lives. This is as true for Christian, Muslim, and Jewish traditions as it is for Hindus, Buddhists, and others. And while it's easy to dismiss faith as not fitting in with a rational view of the world, its influence on how many people see the future, and how their imagined future in turn influences them, remains profound.

But faith goes far beyond organized religion. It seems that, along with our reason, part of our common biological heritage is an ability to believe in what we don't understand or cannot prove. For instance, many people who claim they don't believe in a particular God admit that they believe in some divine being or force, or that they think there's something beyond mere mechanics that determines our future. This tendency even extends to people who describe themselves as rational thinkers, including many scientists.

Whether faith is associated with religious belief, a general sense of there being more to life than we can possibly imagine, or simply a willingness to keep an open mind, most, if not all of us, are to some extent motivated by a vision of the future that isn't proven, or even provable. This may be as simple as having faith in the value of the work we do, or as profound as believing in the importance of treating others with kindness, dignity, and respect. Either way, we all have a remarkable faculty for constructing visions of the future that rely as much on faith as they do reason.

It's almost as if, somewhere along the way, we inherited a set of traits that make us predisposed to believing in futures that we cannot say with certainty are plausible. And, truth be told, the world, and the future we strive for, are all the richer for this, especially as our ability to believe in what we cannot see is deeply intertwined with yet another human trait that transcends mere rationality: our imagination.

20

IMAGINATION

In 1971, John Lennon's seminal anthem "Imagine" hit the US charts. The antithesis of a call to arms (the Vietnam War was in its sixteenth year at the time), Lennon's song exhorts listeners to imagine there is no heaven, no hell, no countries, and no religion—just "people living life in peace."

However you feel about the sentiment, Lennon's song taps into our amazing ability to imagine alternative futures. It's a talent that builds on and catalyzes our ability to learn, to reason, and to have faith in what we don't see and understand.

It's this ability to imagine a future that's different from the present that inspires us to take steps to build what doesn't yet exist. It's at the heart of our faith in a future we don't know, but that we believe could become reality. It's what inspires scientists to seek out new knowledge they imagine exists, and what drives engineers and technologists to transform theories and equations into tangible products they imagine are possible. Imagination is what allows architects to look at blueprints and see towering buildings, inventors to look at a pile of seeming odds and ends and see fantastic machines, and investors to look at a decaying brownfield site and see smart cities and thriving communities. And, of course, imagination is what empowers artists to provide often-transformative glimpses into what might be through their work.

Of course, just as Lennon's "Imagine" is a less-than-perfect recipe for universal peace, our imagination is a less-than-perfect conduit between where we are now and the future we'd like to inherit. Our imagination enables us to be inspired by what might be, but that doesn't mean that everyone's going to be in agreement on what the future should look like and how we should get there. Sadly, there's a dark side to our imagination—torture and genocide are as much a product of it as the desire for universal happiness. And often, our imagination far outstrips what is possible. It's easy, for instance, to imagine the transporters and food replicators of *Star Trek*, or the faster-than-light travel that makes so many science fiction stories possible. But these, as with many imagined futures, lie beyond the limits of what the laws of the universe allow.

Yet, even when it strays into fantasy, our imagination enables us to construct different future possibilities. And, even if they are ultimately unreachable, it can inspire us to open up pathways to futures that, without it, would remain forever undiscovered.

But it's one thing to open up a pathway to the future, and quite another to actually take the initiative to head down it. Our imagination may open the way to the future, but it's our curiosity that leads to us taking the first tentative steps toward it.

21

CURIOSITY

On August 6, 2012, NASA's Curiosity rover landed on Mars. Although this was initially a two-year mission to explore the Gale Crater, Curiosity is still going strong in its exploration of the Red Planet.

Curiosity was aptly named. The rover captured our seemingly limitless urge to ask questions, to overturn metaphorical stones to see what's underneath, and to push the limits of what we know. It epitomizes our thirst for knowledge about what our universe is like. And it continues to capture our collective hearts as it embodies our fascination with the new and the unusual.

As an evolved trait, our curiosity makes us exceptionally resilient as a species. Curiosity is what makes us discontent with what we know, and it's what inspires us to constantly develop new knowledge—something that's vital to surviving in a dynamic environment. It's the "what-if" part of us that is fascinated by what's around the corner and where it'll take us. It's our innate curiosity that leads to us delighting in our ability to map out and understand the world we find ourselves in, and that gives us the wherewithal to navigate our way toward a future that's even better than the present.

Curiosity is what leads us to ask questions that help us connect cause and effect, and that result in new knowledge. It's what inspires to us try new things, just because we can. And it's what compels us ask "how" as we begin to see what is possible in the future, but can't quite envision the way to get there.

Of course, our curiosity can also be a liability—at least for those with more curiosity than sense. Driving blind or tasting random substances, just to see what happens, isn't the best way to approach the future, unless it's a rather abbreviated one you're looking for. And this is where curiosity without intelligence and intellect can be dangerous. Yet, without curiosity, we will never be able to work out how to get from where we are now to where we want to be, no matter how compelling our vision of the future is.

Curiosity alone, though, can't get us to the future we imagine. It prompts us to push beyond what we currently know. But as we push, we need some image, some object in our mind, that hints at what we might find there. And for this, we need to turn to a close cousin of curiosity: our creativity.

22

CREATIVITY

In its 2018 *Future of Jobs Report*, the World Economic Forum stressed the need for employees to become increasingly creative if they're to ride the wave of the future. Emerging technologies and shifting global trends, the report pointed out, are altering the jobs landscape so rapidly that we all need to become more creative in how we make ourselves useful, if we're to stay ahead of the curve. In fact, creativity is now such a buzzword in business circles that it's led some to speculate that "creativity" is the new "innovation."

And yet, while most of us, I suspect, feel that creativity is important when thinking about the future, it's surprisingly hard to pin down what it is exactly.

The roots of the word "create" lie in the Latin *creare*, meaning to make, produce, or procreate. Over the years, it's become associated with the idea of producing something novel that wouldn't otherwise exist, especially when this something is the product of human imagination and ingenuity.

Building on this, creativity can be thought of as developing and using the ability to produce new ideas and objects that have potentially greater value than the raw materials that are used to construct them. I say "potentially" as it's sometimes hard to ascertain the value of creativity, and often what is seen as being worth something by one person is dismissed by another. And yet, even with this uncertainty, there is an essence to creativity that captures the process of taking existing ideas, materials, and products, and forming something novel with them that has the capacity to reveal unique insights and open up new possibilities.

In this way, creativity has the power to help us realize new pathways toward the future. In a shifting global jobs market, for instance, it provides insights into ways in which people can realign themselves with changing needs and expectations. But in the broader context of future-building, creativity is far more than this. It's what enables us to build bridges, step by step, toward the future we aspire to.

It's this nature of creativity that makes it a particularly compelling part of our future-building toolkit. Imagination and faith can help us envisage any number of futures. Knowledge and reason help us sift out those that are plausible from those that are mere fantasy. And curiosity expands our toolkit for building the future we desire as we make new discoveries. But it's our creativity that helps us apply these tools in ways that help get us to where we are going.

In this way, our creativity is hardwired into our future. What happens next week or next year is no longer solely a product of the inevitable flow of time, or the fortuitous fluctuations of natural selection. Rather, it's influenced by our ability to create pathways that both are inspired by how we imagine the future, and lead us to futures that would not occur if the world were left to its own devices.

But creativity is more than opening up tangible pathways to the future. It's also about creating pathways within our minds that lead to imagined futures, and the weaving together of our dreams and aspirations with plausible reality. And nowhere is this more apparent than in the creative arts.

23

ART

Some years ago, while I was working just off Pennsylvania Avenue in Washington, DC, I remember stumbling across a startlingly simple exhibit in the National Gallery of Art. It was a square of white fabric pinned to a wall, and it bore a striking resemblance to the handkerchief I had in my pocket. At the time, I couldn't help wondering what transformed this seemingly insignificant piece of fabric on the wall into art.

Much art, as an act of creativity, holds value that is firmly lodged in the eye of the beholder. My white fabric square (which sadly, I don't recall either the name of, or the artist responsible for) was part of a rich tradition of using common objects in uncommon contexts, to stimulate new ideas and perspectives. Such artistic creativity, while sometimes seemingly trivial, can hold a quite remarkable power to stimulate the imagination of others as it reveals ways of seeing the world that transcend the limits of what is immediately visible. And this extends through time and space to connect our creative understanding of how past, present, and future are intertwined.

Art, in all its forms, is oxygen to our creativity as we contemplate the future. It fuels our ability to imagine alternative futures, and to construct ways of building them. And it does this by enabling us to share ideas, feelings, and beliefs in ways that mere facts and figures cannot convey. Art, in a very real way, enables us to touch the minds of others, and to be, in turn, touched, as we envision and create the future together.

This "artistry" takes on an incredible diversity of forms. Realism within artistic expression helps flesh out visions of future possibilities that are easy to make sense of, and can make tangible what was previously ephemeral. Where artistic expression is more abstract, it forces us to see and experience the world and our place in it differently, by kicking us out of the rut of conventional thinking. And it feeds our imagination by providing us with a tantalizing glimpse of the future, while leaving our minds to fill in the gaps.

Looking back, I'm intrigued that, even though I struggled to make sense of the white square of fabric hanging in the National Gallery that day, the memory stuck with me, and eventually became grist to my thoughts here on the relationship between art and the future. In the same way, art of every conceivable form can act as a surprisingly powerful catalyst for our imagination as we contemplate the future, and a stimulant that feeds and nurtures our own future-oriented creativity.

As art fuels our creativity, it's capable of revealing insights into the consequences of our actions in ways that transcend rational analysis. As it taps into our emotions and instinct, it complements and extends our intellect as we peer into the future. It inspires us to stretch our imagination, and it fills us with hope for what might be. Yet it also reveals the darker side of what possibly lies beyond the veil of the present. And as it does, it has the power to invoke that most visceral of responses: fear.

24

FEAR

In 2003, the writer and actor Max Brooks published *The Zombie Survival Guide*. As a work of creative fiction, it provides a surprisingly detailed description of how to prepare for and stave off a zombie attack. Underneath the apparent fantasy (and Brooks is adamant that the guide was never intended to be anything but what it is—a genuine guide for surviving zombies), the book has a serious backstory. Brooks was terrified by the notion of zombies as a child. As he explained in an interview for *Rolling Stone* in 2017, "The first zombie movie I saw was...an Italian cannibal zombie movie and it terrified the hell out of me." *The Zombie Survival Guide* was, in part, a way for him to reason his way out of his fear.

Brooks's art became both a catharsis for his own fears, and a platform for helping others work through situations where their fear threatened to get the better of them. And while zombies remain firmly relegated to fiction, Brooks cleverly demonstrated how this nexus of fear and art can provide surprising insights into the future. His books are even used these days to teach military strategy.

Fear is a complex mix of instinct, rational thought, and physiological response. It can entail a very physical reaction to an instinctive threat—a quickening pulse, profuse sweating, a churning stomach, and heightened levels of anxiety. It's part of the fight-or-flight response we evolved in the face of other animals that would as soon eat us as pass us by. Yet our fear response is also part of our learned experience. And so our body—bizarrely—reacts to anxiety over a particularly tough exam, or the thought of being caught in a lie, or even being thrown into a socially stressful situation, in the same way it would if faced with a ravenous beast. And in each case, part of our brain is constructing a vision of a future associated with pain and suffering, and is desperately working out how to avoid it.

Fear, it turns out, is an extremely strong biological thread that connects future possibilities with present actions. As we envision the future—whether through gut instinct, or logic and reason—it's fear of what might occur that motivates us to try to change that future. Hundreds of thousands of years ago, the two most basic tools we had at our disposal to do this were to stand our ground, or to run away. These days, we've substantially expanded our fear-response toolkit.

Living in the modern world, we have a vast array of personal, social, and technological ways in which we can fight fear, avoid it, or simply make it disappear. Yet at the heart of all of them is our ability to foresee and to viscerally imagine the consequences of futures that we'd rather avoid. Fear connects us emotionally with the future, and scares us into doing what it takes to avoid those that stress us out.

But why is fear of the future such a powerful motivator? Why does it so often take more than just an intellectual understanding of what the future might hold to galvanize us into action?

Part of the answer lies in the reason why we experience fear in the first place. And much of this comes down to the possibility of losing something that is deeply precious to us, whether that's our identity, our health, our life, or that most devastating of situations, losing someone we love.

25

LOSS

Sandy Peckinpah was blessed with three beautiful children. Then one day, the unthinkable happened. As she wrote in 2016, "... without warning, my life changed. My beautiful 16-year old son came home from school complaining of a headache and a fever. The doctor diagnosed him with the flu. But it wasn't. Sometime during the night, my boy was taken from me forever. I found him the next morning in his bed, lifeless. The misdiagnosis was actually a swift and deadly form of bacterial meningitis."

Sandy's experience is, tragically, repeated day after day the world over, as parents face one of the worst tragedies imaginable: the loss of a child, and the future they should have lived.

Loss—especially loss of life—is often so painfully entwined in our sense of the future that it is difficult to write about. And yet, our fear of losing someone or something precious to us, and the grief that accompanies it, or our experience of such a loss, deeply affects our relationship with the future. As parents, we dream of the amazing futures that our children have the potential to weave, and we dread the imagined scenarios that lead us down dark, painful future pathways. Even if we haven't suffered the loss of a child, the mere possibility of something so unthinkable happening is enough to hurt as physically as a punch to the stomach. And of course, it's not just the loss of a child. The loss—whether actual or feared—of any loved one hurts deeply, and it hurts badly.

Loss, together with the imagined possibility of loss, ties us with a near-unbreakable cord to the future. The threat of losing something that's near and dear to us drives us to circle the metaphorical wagons as we strive to protect what we have. And it's what so often prompts us to build a future that we believe will be safer and more secure than the present.

As a result, we strive to create a future where our children are happy and healthy, where our loved ones thrive, where there is no pain, hunger, poverty, or injustice, and where we live in balance with the world that sustains us. It's our fear of loss of what we have, and of what could be, that so often galvanizes us to get up and change the future, rather than simply being bystanders to what unfolds.

But what if our fear of loss hits a brick wall? What if we can see no way to navigate it, and no pathway forward to the future we desire? When this happens, our capacity for imagining a future that is better than the present, but which lies beyond our grasp, becomes a very real threat to what we value—an inevitable loss that we are powerless to prevent. And with this, we risk slipping into despair.

26

DESPAIR

Despair has a dark and complex relationship with the future. It's a feeling or state that only exists if we can imagine a future that's preferable to the present, but that we have no conceivable way of reaching. Despair is a state of being trapped in a painful present while being aware of a future that will always remain beyond us. It's living in a world where time and opportunity have become ossified, and we know it.

If we had no memory, no imagination, no curiosity, and no faith, we would be immune to despair. Because we couldn't imagine anything other than the present we exist in, we would be blissfully unaware of what could be, and resigned to live with what we have.

Yet this isn't our world. The gift of being able to imagine the future and to potentially change it also comes with the curse of living with the reality that there are futures we can perceive, but which lie beyond our control. Most of the time, we live with this by focusing on what we do have influence over, and reconciling ourselves to that which we don't. We have a remarkable ability to put our visions of the future into categories that represent the attainable, the aspirational, and the unreachable. We work hard to achieve the first, and stretch ourselves to bring about the second. And the third is usually relegated to our dreams and fantasies.

But what if the only category with anything in it was the third one? When we have no attainable or aspirational visions of the future, we lose our sense of agency over where we're going. And as the thread between where we are now and the future we long for is severed, we're cut adrift. Because we are creatures of time, moving irrevocably from past to future, we suffer the cruelest fate of being able to imagine a better world, but being powerless to reach it.

Sadly, this is a fate that's been imposed on a shocking number of communities over the centuries, as our baser instincts lead us to marginalize, abuse and repress others. We may be architects of the future, but we're also disturbingly good at stealing the futures of others when it suits us. Or worse, imprisoning them within the walls of futures we impose on them.

Yet, even within the seeming pit of despair that our ability to imagine the future sometimes leads to, there are often hidden possibilities. Fortunately, the future is rarely as simple as we imagine it to be, and the pathways toward it are often more numerous than we realize.

27

POSSIBILITY

There's a theory that, every time a decision is made, the universe splits in two, with each resulting parallel universe representing a different possible outcome. And so, the theory goes, for everything that could conceivably happen, there's a universe where it has.

It's an intriguing idea that keeps theoretical physicists occupied, although in practical terms it gets itself tied up in knots pretty fast. Yet despite its seeming absurdity, it does underline the reality that we live in a universe of possibility, one in which the pathways between where we stand in the present and where we end up in the future are far more numerous and varied than we often imagine.

Possibility is part of the landscape we travel through as we journey toward the future. Without possibility, the future becomes fixed, and we're back to despair. Fortunately, even without invoking theories that depend on infinite parallel universes, the cosmos is in a constant state of flux. And with this comes a bewildering, if not infinite, array of possibilities.

From the oscillation of a fundamental particle and the passing of a ray of light, to the unzipping of a strand of DNA, the universe we inhabit is constantly changing, and with each change, nascent possibilities hang on the edge of becoming reality. These minuscule ticks of the cosmic clock coalesce into larger waves of possibility, which in turn can carry us forward in strange and unexpected directions. Through the unlikeliest of confluences, they reveal hitherto unseen opportunities; sometimes to ourselves, and sometimes to others who more clearly see the pathways to a better future that lie in front of us, and are able to take us by the hand and guide us there.

These possibilities chip away at despair as they reconnect us with the ability to forge our own path forward. And as they do, they lead to hope.

28

HOPE

On April 24, 1916, one of the most perilous journeys in human history began. In a last-ditch attempt at survival, an intrepid band of explorers stranded in Antarctica boarded a small, barely-seaworthy boat, and began the perilous journey to the isolated island of South Georgia.

Nearly two years previously, the British explorer Ernest Shackleton had set out to lead a group of explorers in the first traverse of the South Pole. But before they got there, disaster struck as their ship was trapped in sea ice and began to disintegrate. Over the following months, the crew struggled to survive on the ice, and in early 1916 they made their way to a small island at the edge of the Weddell Sea where, faced with bitter cold and dwindling food supplies, their plight began to look increasingly hopeless.

In a final bid to escape the clutches of the Antarctic, Shackleton led a crew of three in one of the original ship's lifeboats across eight hundred miles of one of the toughest ocean crossings in the world, in an attempt to reach a small whaling community on the island of South Georgia.

On May 10, 1916, the three explorers landed on the island, and eight days later, they successfully hiked into the whaling community at Stromness after one of the most incredible journeys in recent history. Against all odds, Shackleton envisioned and engineered a future in which his crew were returned to safety. By all accounts, a sea crossing of eight hundred miles at that latitude in a small lifeboat, with only the most basic of navigation aids, while attempting to reach a small, mountainous island, was foolhardy at best. And yet Shackleton had hope. Driven by the alternative of losing his crew, one by one, to lack of food, illness, and the elements, he took a calculated risk. And it paid off.

Shackleton's story stands out as an impressive feat of survival, but it's far from the only one. And what each has in common is the ability of people to defy the odds as they peer into the future and hope for the best.

Of course, hope on its own can do little to change the future. Yet, with all of the future-oriented skills and tools we've inherited, humans are exceptionally good at teasing out the fine threads of hope from the jaws of despair, and finding ways to futures that might, at first glance, seem unreachable.

Whether this ability to hope and succeed is simply part of our evolutionary heritage, or is some emergent property that transcends our biological roots, it's hard to tell. But what is certain is that we have an amazing ability to see beyond what seems possible, and to inspire others with the hope of a better future. And in part, this ability relies on the uncanny skill with which we weave stories that connect past, present, and future, together.

29

STORIES

I was twelve years old when the first *Star Wars* movie was released back in 1977. I didn't see it when it first came out, but I do remember my French teacher—who was not prone to waxing lyrical about science fiction films—being barely able to contain herself as she told us the wonderful story of beautiful princesses, dashing heroes, and evil foes.

We all love a good story. It's in our blood—and for good reason. Stories are how we make sense of the world around us. They help us piece together the convoluted puzzle that connects cause and effect, and they enable us to map out our lives, and those of others, in relation to the future. More than this, stories allow us to share our visions of the future with others, and provide them with a glimpse of the world as we see and experience it. And because of this, stories have become part of the social fabric that holds us together as we journey into what's to come.

Shackleton was able to chart a path to safety because he told himself a story of how to get from a situation that seemed desperate to one that was not. But it was his ability to convincingly tell this story to others that inspired them to follow him.

Star Wars isn't quite in the same league as Shackleton's epic real-life adventure. But it still taps into our innate ability to create and share stories that inspire and motivate us. And more often than not, the most powerful stories we weave are those that involve the future.

Star Wars thrusts us into a story that occurred "a long time ago in a galaxy far, far away." It's a tale that revolves around someone else's fictional future that hangs in the balance, as the forces of good and evil fight it out. Yet despite the locational and temporal displacement that comes with such a fictional story, it resonates with our own dreams and hopes. And this is where story-telling begins to weave its magic. Even when the stories we tell each other take place in made-up universes, they can inspire us to see our relationship with the future in a new light, and to transform what we imagine to be possible into tangible reality.

When contemplating the future, we are profoundly affected by what we stand to lose, and what we hope is possible, and we have an astounding capacity for building creative, plausible visions of what the future might look like. Yet it's only when we begin telling ourselves stories of how we might get to these envisioned and imagined futures that they begin to crystallize.

These stories thread together fact and fiction as we mentally map out the path between where we are now and where we want to be. Some are grounded in practical reality, such as the story that Shackleton crafted around how he was going to ensure the survival of his crew. Others are caught up in a world of fantasy that's as implausible as the *Star Wars* universe. Yet every story we tell ourselves illuminates, in some part, pathways to the future.

In this way, stories become the pivot point between being able to imagine the future and beginning to build it. They set the stage for the journey and help form the mental maps that chart a way between past, present, and future. But they don't necessarily provide the means to get there.

For this, we need to turn to another quite amazing ability that humanity has developed and nurtured over thousands of years: our capacity for invention.

30

INVENTION

Google "weird inventions" and it quickly becomes apparent that, as a species, we have a near infinite capacity to fill perceived holes in our imagined futures with the oddest of devices. Whether we're talking about irons that double as coffee mugs, or pizza scissors, it's quite astounding what some people feel we can't live without.

Many of these inventions come across as the products of a warped imagination. But despite this, they reflect our ability to not only spin stories about a future that's different from the present, but also populate it with all manner of objects that don't yet exist.

Invention is an essential part of our ability to problem-solve by building bridges between the present and the future. Some of these inventions may seem a little frivolous—I'm not sure that a future without pizza-scissors is quite the tragedy that their inventor might imagine it is. But they all contribute to a rich and varied "primordial soup" of inventiveness.

Just as genetic mutations provide the variation that enables organisms to adapt to the future through natural selection, our propensity to invent provides the material variation that enables us to build tangible pathways to the future in an ever-changing world. And while most inventions will ultimately fall by the wayside, the right invention at the right time can be a game-changer.

As a result, our world is built on a deep and wide foundation of invention. The wheel, the ways we use fire, bricks, ball bearings, pens, bagels, espresso, cars, laptops...our modern lives are a living record of a past in which someone had the wherewithal to ask "what if" and imagine an invention-shaped hole in the future that they felt compelled to fill.

This capacity for invention is a natural extension of our sense of the future and our ability to tell ourselves stories about what it could be like. It represents a leap of faith in which we believe we can create a future that currently only exists in our mind. And just like genetic mutation, it generates a magnificent array of possibilities.

Invention is what propelled humans from a hunter-gatherer existence, ten thousand years ago, to becoming farmers and city dwellers. It fueled the Enlightenment, as scientists invented new tools to explore the world around them and new ways of sharing their ideas. And it powered the Industrial Revolution as we learned how to harness the power of water and steam. Invention gave us the green revolution, the digital revolution, and the social media revolution, and it continues to push us into an increasingly complex future at an ever-increasing rate.

And yet, invention on its own is a rather inefficient engine of change. Like random genetic mutation, invention leads to problem-solving options and possibilities that ensure we are resilient to change. But like natural selection, it's slow. If invention is to help us create the future we desire as fast as we desire it, it needs a speed boost. And this sets the stage for the close cousin of invention: innovation.

31

INNOVATION

In 2003, *Wired* magazine ranked "innovation" as the most important and overused word in America. Little did the magazine's editors realize that they hadn't seen anything yet!

There are times when it seems innovation is all that matters in the modern world. No matter how smoothly or poorly things are working, we're told we have to innovate. Students are taught to be innovators. Businesses are informed they need to innovate or perish. Governments are wringing their metaphorical hands out of fear that they are not being innovative enough. Even artists and designers are being informed they have to innovate more, as if their previous creative efforts weren't sufficient.

And yet, when pushed, few people are able to articulate what innovation actually is.

All too often, innovation is taken to mean change, or doing something differently. Yet, if the aim is to build a better future as fast as we can, this is a really bad idea. It's akin to throwing out the map at the start of a journey, and prioritizing speed over direction.

Rather, just as a successful journey means moving in the right direction toward a specific goal, true innovation involves coming to grips with where you are heading and why, as well as any change that this entails along the way.

One of the most useful definitions of innovation I've found, and the one I use with my students, is the translation of creative ideas into products and processes which provide sufficient value to others that they are willing to invest in them. It's a way of thinking about innovation that reflects the importance of creativity and change, but only when they lead to something that enables someone to achieve what's important enough to them that they're willing to invest their time, money, or effort in it.

It's innovation like this—focused, targeted, purposeful change, rather than undisciplined creativity and undirected invention—that is driving some of the most transformative trends in technology today. Advances in computing and data processing, the emergence of machine learning systems and artificial intelligence, gene editing, personalized medicine, smart cities, and many other trends are all a direct result of our ability to harness our creativity and inventiveness in service of building a future we imagine is possible.

Of course, innovation is a double-edged sword, and for every advance we make, untold numbers of unintended consequences begin to ripple through society and out toward future generations. Yet, when we get it right, innovation is a powerful part of our future-building portfolio. It's not the only one, though. The most successful innovations are those that fit into a bigger picture of how we imagine the future could be. And, increasingly, this is a picture that is guided and governed by the principles of design.

32

—

DESIGN

On January 24, 1984, Apple Computer cofounder Steve Jobs stepped out onto a stage at De Anza College, California, and changed history. The event was the much-anticipated launch of the iconic Macintosh 128k personal computer.

Surfing the wave of one of the most memorable ad campaigns ever, the Macintosh masterfully blended together innovation and design. And in doing so, it changed the future of personal computers. But there was a small appendage attached to it which, if anything, has had an even greater impact on how we use our electronic devices: the computer mouse.

These days, touchpads and the touch screens on smartphones and tablets are beginning to supersede the mouse. But between 1984 and the early 2000s, the mouse ruled the computer interface roost. And its success was due in part to a particular blend of innovation and design that has become known as "design thinking."

Design thinking places humans at the center of an iterative process of rapid idea generation and testing, and it's found its way into everything from the design of disposable cups to how advanced technologies are governed. Its origins go back decades, but the concept was popularized in part by two Palo Alto designers, Dean Hovey and David Kelly.

In 1991, Hovey and Kelley founded *IDEO*, one of the world's most successful and influential design companies, and one that effectively put design thinking on the map. But even before the launch of *IDEO*, Hovey and Kelly were working together on innovative designs. And perhaps the most influential of these was the mouse that shared the stage with Steve Jobs, back in 1984.

The story of the Apple mouse illustrates just how deeply design, innovation, and the future are intertwined. You can't innovate your way to the future without designing something new that's of value to someone else. Design by intent, rather than by accident, substantially amplifies our ability to change the future in ways that are potentially useful. More than this, design thinking brings a very human element to innovation. It encourages innovators to think about not only *what* they can build, but also *how* it enhances or otherwise affects the lives of others.

This human-oriented side of design helps increase the value of innovation to the people who use it, and whose lives are affected by it. And it becomes especially important when developing powerful technologies that could, if not handled sensitively, cause more harm than good in the future. But there's also a humility to design thinking—at least in some ways that it's practiced—that recognizes the unpredictability and ultimate uncontrollability of the future.

This is seen in the way design thinking and the complex challenges associated with "wicked problems" come together. Horst Rittel and Melvin M. Webber coined the term *wicked problems* in 1973, in recognition of just how unpredictable the future is. These are problems that are so slippery that the very nature of the problem changes as attempts are made to solve it. Wicked problems are essentially unsolvable problems, where the best that can be done is to develop temporary solutions on the path toward ever-shifting goals, and this in turn requires a uniquely fluid approach to design.

Rittel and Webber's work focused on social policy planning. But since they popularized the idea, it's become increasingly clear that much of the world we live in is dominated by wicked problems—especially where people are involved.

Design thinking recognizes this "wickedness" in its approach to building the future. It's a highly pragmatic process that reflects the near-impossibility of getting things just right where people are involved, and instead encourages ways of building pathways to the future that are right enough for the moment, yet agile enough to adapt to a changing landscape.

Through the smart use of innovation and the savvy application of design approaches, we are more adept now than at any previous point in our history in imagining what the future might be like, and how we might get there. And because of this, we have within our grasp the ability to dramatically change the future.

How this plays out, of course, depends on whether we have the maturity to use such profound capabilities responsibly. Yet without a doubt, we are on the edge of being able to transform the future in ways that we could only dream of a few short decades ago.

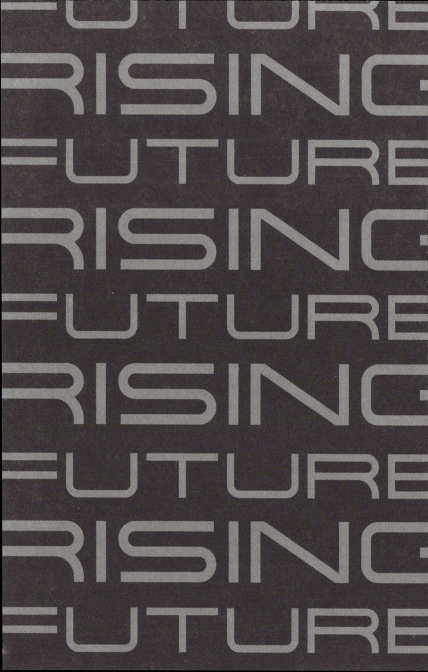

PART 3

BUILDING THE FUTURE

"Tomorrow belongs to those who can hear it coming."

—David Bowie

33

TRANSFORMATION

Barely a hundred years ago, the pace of change around the world was so slow that, for many communities, life was little altered from that experienced by previous generations. Rural communities were caught up in a rhythm of life that had persisted for hundreds, if not thousands, of years. And where there was innovation, it caught on slowly. Then, in the early 1900s, everything changed.

In 1959, the British author Laurie Lee poignantly captured this transition as he experienced it as a young boy growing up in a sleepy Cotswold village. "The last days of my childhood were the last days of the village," he wrote in his autobiography *Cider with Rosie.* "I belonged to that generation which saw, by chance, the end of a thousand years' life."

It's been many decades since I first read those lines, but they still stick with me. They mark the beginning of a quite remarkable transformation in the lives of many people that was brought about through modern science and technology. And they emphasize just how radically modern innovation has altered our perspective on the future.

As Lee writes, a hundred years ago "the horse was king, and almost everything grew around him: fodder, smithies, stables, paddocks, distances, and the rhythm of our days. His eight miles an hour was the limit of our movement, as it had been since the days of the Romans. That eight miles an hour was life and death, the size of our world, our prison."

Since then, our technologies have expanded our world to the farthest reaches of the solar system and beyond. It took a thousand years for the car to supersede the horse, and a mere half a century for us to progress from crude vehicles to space rockets. Digital technologies now enable us to beam in remotely to nearly anywhere on Earth, and even beyond. We're sending probes to the farthest edges of the solar system, we're planning to send humans to Mars, and we're observing the births and deaths of stars in galaxies far beyond our own. In just a few short generations, technology innovation has altered our slow, steady march toward the future into an ever-accelerating race to achieve possibilities that, until recently, we had no conception of. And as it has, it's utterly transformed our perceptions of what lies ahead of us.

Over the past decade alone, trends in technology innovation have emerged that have elevated what was once science fiction into science-possibility. And in doing so, they've turned our idea of what we can achieve on its head. Advances in gene editing are opening up the secrets of precisely reprogramming DNA. Artificial intelligence is leading to machines that can outperform humans in areas we previously thought were sacrosanct. And nanotechnology is giving us unimaginable mastery over the atoms and molecules that everything around us is made of.

As this pace of innovation accelerates, our vision of the future is changing as fast as the technologies that promise to crystallize our dreams into reality. As a result, we're building a powerful feedback loop between what we can imagine and what we can achieve. And it's one that is transforming the world around us as we watch.

This is heady stuff. But with these accelerating advances in technology there's a danger of overreach, as we begin to imagine and be influenced by futures that, despite our dreams and aspirations, lie far beyond what's possible.

34

ACCELERATION

There's a clever trick that some technological visionaries use to predict the future. They chart the rate of change in a specific area over time—computing power, say, or DNA sequencing. But they do this using powers of ten, so that every small vertical increment on the charts they proudly flourish represents a tenfold increase in capability. As a result, unimaginably large leaps of imagined possibilities are compressed into seemingly smooth and inevitable trajectories.

These selfsame charts are then used to predict how long it'll take before we hit truly transformative tipping points in our technological future—the point, for instance, at which whole-genome sequencing is faster than typing a password into your phone, or the decade within which we'll be able to upload the human brain into a supercomputer. Or even the point at which we'll have the knowledge and tools to cure presently incurable diseases. And because there's such a seemingly small difference between a tenfold and a millionfold projected increase in what we might achieve on these charts, the predictions are both compelling and misleading.

These are exponential growth predictions. Rather than technological capabilities increasing at a steady rate, they assume that our ability to change the future through technology will continue to accelerate. And to some degree, they are correct—at least in part. There are many areas where the speed of innovation is still on the increase—computational power and gene editing are just two. And yet, exponential extrapolation into the future, based on an assumption of continued technological acceleration, can be problematic.

The classic example of such growth is Moore's Law. In 1965, Gordon Moore, the cofounder of Intel, projected that the number of components that could be engineered into an integrated circuit would continue to double each year, leading to an associated exponential increase in computing power.

Moore's projection turned out to be roughly right, although it was buoyed by the semiconductor industry working hard to make sure his vision of the future became reality. Recently, the rate of increase has slowed, as individual electronic components reach the limit of what's physically possible. But Moore's "Law" has become so engrained in some people's thinking that they fervently believe that new technologies will continue to enable a doubling of computing capabilities every year, despite the physical limitations we face.

This, unfortunately, is fanciful thinking. Exponential growth is a seductive way of viewing the future. It makes it feel that almost anything is possible given time—including uploading ourselves into cyberspace, or believing it's worth cryogenically preserving ourselves (or our children) until we can be revived and refitted with superhuman bodies. Yet faith in exponential predictions ignores the reality that the pathway between present and future is neither linear nor exponential. Rather, it is deeply convoluted, with hurdles and pitfalls emerging where we least expect them, and new opportunities serendipitously emerging that confound all expectations. And even where capabilities continue to accelerate, the direction in which they will continue this trajectory is often far from clear.

And yet, the promise of exponential growth continues to inspire many to strive for a future that is radically different from the present. And in particular, it lies at the heart of the transhumanist movement: a movement that's committed to building a future that transcends our biological roots.

35

TRANSCENDENCE

For billions of years, the future of organic life has been dictated by the laws of natural selection and the constraints of its biological foundations. Because of this, if a microbe sitting in someone's gut, or hanging out in a fetid pond, could imagine its future, it would be one that lay within a narrow range of possibilities—there are only so many things a microbe can do within the biology it inherits.

Humans, of course, have advanced far beyond the biological limitations that constrain microbes. We have intelligence, reason, creativity, and the capacity to imagine futures that far exceed what exists in the present. And yet, we are still shackled by our biological heritage. We're born, we live a handful of years, we get old, and we die. Our existence is wrapped up in a tight bubble of possibility that determines the limits of how strong we are, how fast we can travel, how smart we can become, and how vulnerable we are to the environment we live in. We can dream big when it comes to the future, but our dreams are still neutered by our natural-born capacity.

To truly *live* the future we can see in our mind's eye, we would need to find a way to break free of the biological bubble that currently limits us, and transcend our evolutionary heritage.

Innovation, of course, is one way we've discovered to push the boundaries of this bubble. We imagine traveling faster than our legs will carry us, and we invent the bicycle, followed closely by the car. We dream of flying like birds, and we invent planes. We speculate about visiting other planets, and we invent spaceships. Every day, technology innovation is transforming our dreams into reality, and pushing the bounds of what is possible. And yet, at the heart of all of our inventiveness is the same soft, weak body, infused with hormones, hobbled by limited intelligence, and saddled with a shockingly short lifespan.

What, though, if we could use our creativity and inventiveness to overcome our biological limitations, even to the point of cheating death itself?

In 2014, the futurist Zoltan Istvan announced that he was preparing to run in the 2016 US presidential campaign. What set him apart from every other potential candidate was his rejection of the inevitability of death.

Istvan is a self-proclaimed "transhumanist." As such, he believes that science and technology will, one day, free us from the tyranny of biology and enable us to live forever. He even goes so far as to imagine a future where biology itself is outmoded, and every part of us is replaced by machines, including our brain.

Istvan's 2016 presidential run never got off the ground. But his vision of the future persists, and it's one that's shared by many others, including a number of leading academics and entrepreneurs. But just how realistic is it?

The short answer is, sadly, not very.

The less we know about how the human body works, the easier it is to imagine replacing parts of it with machines that make it work better. This is great for speculating about imaginary futures, but it's problematic when we come to building them. It's easy to imagine replacing limbs with robotic arms, or swapping out our rather messy digestive system with a high-tech power source, or even augmenting our slow, mushy brain with the latest in AI microprocessors. These are the dreams of science fiction writers and transhumanists. Yet the more we learn about our bodies and their underlying biology, the more we realize just how far we are from getting close to what over three billion years of evolution has, so far, achieved.

We may think of ourselves as limited, vulnerable, messy, and short-lived. But as scientists are still learning, our bodies are sophisticated super-machines that we've barely begun to make sense of. And to imagine we could improve on what we don't yet understand verges on the naive.

Yet our dreams of a future where we are smarter, stronger, and live longer, have a profound influence on our visions of what's to come. Despite the seeming naivety of transhumanists, their vision is already inspiring innovators to push the boundaries of what is possible. And as they do, they are actively transforming the future, to the extent that progress in developing technologies from robotic prosthetics and 3-D-printed organs to brain-computer interfaces and many others, is accelerating far faster than most people would have predicted a few years ago.

Despite my skepticism over the desirability and achievability of what transhumanists aspire to, it's hard not to be impressed by the accelerating rate of change their ideas are associated with. And it's intriguing to ask where this might lead as we strive ever harder to create a technology-driven future. This brings us to a vision of the future that lies at the heart of many transhumanists' imaginations, and it's one that predicts we're heading toward a point beyond which it is impossible to predict what happens next. This is a point in our collective future that is often referred to as the "singularity."

36

SINGULARITY

In 2006, the engineer, futurist, and transhumanist Ray Kurzweil published the bestselling book *The Singularity is Near*. Kurzweil wasn't the first person to describe what he saw as an impending technological pivot point in humanity's future, but he was highly influential in popularizing the idea.

Kurzweil deeply believed that it was only a matter of time before humans transcended their biological origins. And by studying exponential growth curves across technologies ranging from computing to biotechnology, he estimated when this was likely to happen—which turned out to be far sooner than most people expected.

According to Kurzweil's estimations, in four or five decades from now, we'll reach a point where artificial intelligence is so powerful that machines will be able to invent and construct even more powerful versions of themselves. These, in turn, will design and build even more powerful intelligent machines, and do it far faster than their predecessors, until we get to the point where there is an almost overnight "explosion" of AI capability that leaves human intelligence standing in its wake.

This is the point that transhumanists and others refer to as the singularity. And they do so for the simple reason that, just as it's impossible to see beyond the event horizon around a physical "singularity" created by a black hole, they claim that it's equally impossible to see what lies on the other side of this technological singularity.

If a technological tipping point like this were to occur, all bets would be off when it comes to predicting the future. Our minds, our foresight, and our billions of years of evolved instinct, intellect, and inventiveness have done little to equip us to see beyond such an acceleration in technological advancement. And because of this, we are blind to the possibilities that lie beyond it.

Fortunately, the very idea of a technological singularity is sufficiently speculative that there is plenty of room to imagine a future where it does *not* occur. It's a fascinating intellectual idea to play around with. But there are too many tenuous assumptions built into the thinking behind the singularity to make it much more than an intriguing but exceptionally low-likelihood possibility.

This allows us to breathe a collective sigh of relief, and get on with the more mundane task of peering around the corner of time as we consider more plausible futures. But before we leave the singularity, there is one aspect of it that does need to be addressed, and that's the possibility that we have already experienced it—albeit in blissful ignorance—and we're all nothing more than virtual players in a stupendously complex post-singularity computer simulation.

37

SIMULATION

It's intriguing to imagine the possibility that, at some point in our future, we'll be able to simulate every last ounce of our lived experience so convincingly that it becomes impossible to separate virtual reality from the real thing. Some thinkers—including Elon Musk and the late Stephen Hawking—believe that, someday, this will be possible. Others believe it's already happened.

The theory that we are all living in a vast computer simulation goes hand in hand with transhumanism and the singularity. Believers claim that we're getting so good at replicating reality in cyberspace that it's only a matter of time until the virtual realities we create become better than real life. And from there, it's just one small step to convincing yourself that, at some point in the past, someone created computers powerful enough to make this happen, trapping us all in some glorified computer game, with a made-up past and a predestined future.

There are, as you might imagine, one or two flaws to this logic. For one, to simulate the universe at a level of detail that was indistinguishable from the real thing, you'd need more power and resources than currently exist in reality—

or so the argument goes. Even if this were possible, though, there's another conundrum we face: If we're living in a simulation that's so good as to be indistinguishable from physical reality, does it matter?

It probably would, if we could somehow hack the underlying code and change our experience of reality. This is one of the dreams, or possibly the nightmares, of the simulation scenario. And it's where the possibility of everything around us being little more than computer code has the potential to change how we think about the future, and even challenge how we grapple with the very nature of the future.

If everything we know and experience is determined by code being run on some celestial computer, surely it should be possible to rewrite the code and alter the rules that define our lives. In such a future, why would living forever or gaining superhuman powers require advanced technology when a simple code change could achieve the same ends? The same goes for adjusting the length of the day, or traveling faster than the speed of light. If we're all in a simulation, what's to stop us hacking our way to interstellar travel, matter transporters, and food replicators—all while eliminating poverty and disease at the same time?

This is how the theory that we're living in a simulation begins to worm its way into how we think about the future. And yet, the idea is perilous—not least because it raises false hopes and encourages naive ideas. Perhaps the biggest issue, however, is coming to terms with the constraints of self-consistency in such a sophisticated simulation.

In simple simulations, the laws of physics can easily be thrown out the window. Where the internal laws lead to inconsistencies or glitches, they can be glossed over relatively easily. The simulation ends up being crude and unrealistic, but it works.

But the more complex a simulation gets, the more internal consistency is needed to ensure it doesn't crash or grind to a halt. Basic laws of behavior need to be laid down, much like the laws of physics, and bending or breaking them becomes increasingly hard as the effects of interventions ripple through this virtual reality. Eventually, you get to the point where the processing power required to bend the underlying laws even slightly without the whole system crashing is so immense that it's untenable.

At this point, the simulation becomes indistinguishable from reality, even down to the immutability of the laws that underpin it. As a result, even if we were living in a perfect simulation, we'd never be able to tell.

In effect, while it's entertaining to imagine that we're living in a simulation where, if only we knew how to hack the system, we could change the future, this type of thinking is probably best left to the world of science fiction. And yet, one of the spinoffs of this train of thought that does merit attention is the idea of hacking the future—not in terms of rewriting the code governing some hypothetical simulation, but by rewriting the very real "codes" that tie the past and the future of our surroundings together.

38

HACK

In 1975, Raphael Finkel—then a PhD student in the Stanford Computer Science Department—published the inaugural version of the *Jargon File*. The *File* was a collection of hacker jargon from tech communities at MIT, Stanford, and elsewhere, and as it's evolved, it has become the unofficial bible of hacker culture.

Hacking often conjures up images of malicious coders trying to throw a digital wrench into computer systems. We're regaled with stories of hackers launching distributed-denial-of-service attacks that take down computer networks, or trying to influence elections, or infecting personal computers with viruses and other malware. Yet these do a disservice to a broader community of hackers who are fascinated by novel and often playful ways of altering the future, just because they can.

The current version of the *Jargon File* lists eight definitions of "hacker," with only one of these including malicious intent. According to the *File*, a hacker is someone who enjoys programming (coding in today's parlance), appreciates problem-solving for its own sake, is a fast programmer, is an enthusiast, and enjoys the "intellectual challenge of creatively overcoming or circumventing limitations."

Increasingly, this notion of hacking is being used well beyond the confines of coding. Hackathons are hosted at universities and other sites where groups of students and others work around the clock to "hack" challenges as diverse as developing-next generation robots to combatting climate change. Do-it-yourself biologists are learning how to "hack" living organisms. Transhumanists are "hacking" their own bodies with a variety of technologies. And more and more people are experimenting with lifestyle "hacks."

Increasingly, the spirit of hacking is being applied to any system where, by playing around with it, it's possible to make it work faster, better, or just differently. And underlying these trends is the idea that the future can be altered by changing the underlying "code" that defines it, whether this code consists of the ones and zeroes of digital computers, the DNA of biological organisms, or a myriad other complex, intertwined threads of cause and effect that form pathways between the present and the future.

In effect, modern-day hacking is about recoding the present in order to redesign the future. It's a trend that reflects a growing confidence in our ability to take control of where the future is going, and our ability to steer it where we want. And with the advent of technologies like gene editing, nanotechnology, AI, and even behavioral science, the hacker's toolkit is expanding rapidly.

With an internet connection, some basic resources, a sense of adventure, and plenty of patience, it's easier now than ever before for people with no formal training to become future-hackers. This, I must confess, is something that I find liberating, as more and more people are able to hack their own future and the futures of those around them. And yet, this growing trend also raises tough questions around where the boundaries lie between what is a responsible hack and what is not. And what makes these questions more difficult still are the challenges of working within a deeply interconnected and complex world.

39

COMPLEXITY

On July 31, 2012, one of the largest blackouts in history swept across North India. It started as a single power line in the state of Madhya Pradesh became overloaded and tripped. Within minutes, the disruption had spread through the electrical grid, as system after system struggled and failed to pick up the slack. In the ensuing chaos, over six hundred million people found themselves temporarily facing a future without power.

While power outages are not uncommon in India, the sheer number of people affected by the 2012 blackout is a testament to the unpredictability of complex systems such as power supply grids. As with many such networks, North India's system depends on maintaining a fine balancing act between supply and demand, as unpredictable expectations push a complex web of generating stations, transmission lines, and distribution networks to the limit. And, like all complex systems, the line between normal operations and failure is devilishly hard to spot.

Unlike the predictability of, say, tossing a ball, complex systems are defined in part by their unpredictability. Complexity tangles the threads between cause and effect to such an extent that some future effects simply cannot be predicted, even if we know everything we think we need to know about the causes we believe contribute to them. Despite the future being the result of cause and effect, complexity conspires to obscure our view of the pathways between them.

For a species that has learned to transcend its evolutionary roots through mastering the links between cause and effect, this is a rude awakening. It goes against everything we've come to understand about how we're connected to the future, and how we can design and engineer it.

It was the mathematician and meteorologist Edward Lorenz who provided some of the earliest insights into complex systems and their sometimes chaotic nature. Lorenz was fascinated with predicting the weather, and fervently believed that, with enough data, this should be possible. After all, it was just physics. Yet the more he tried to model trends in weather patterns, the more he realized that seemingly insignificant variations in his starting conditions led to profoundly different outcomes. Try as he might, he simply could not draw a straight line between initial causes and final effects.

This apparent disconnect is popularly known as the butterfly effect, after the idea that the flapping of a butterfly's wings on one side of the world can lead to a chain of events that results in a tornado on the other. This is something of an exaggeration, but we now know that the more complex a dynamic system is, the harder it becomes to predict the outcomes of seemingly insignificant chains of events. And at some point, it becomes impossible to predict what the future will look like, despite our best efforts to design, engineer, or hack it.

Of course, there remains a lot that we can predict. We can predict where and when the sun will rise, and the flow of the seasons. We're able to predict the likely outcomes of unhealthy eating and drinking habits, or the consequences of not vaccinating large numbers of people. And we can predict—in broad terms at least—how human activities are adversely impacting the environment. These are all possible because of our unique combination of reason, imagination, and inventiveness. But it's in the details that the devil of complexity hides. The more precise we try to be with our predictions of the future, the less likely they are to be accurate. And in some cases, irrespective of how much we think we know, complexity results in a future that utterly confounds us.

The danger, of course, is that we become so enamored with our brilliance that we choose to overlook this, and act as if the future is something we can fully control.

40

HUBRIS

In 2003, the then-director of the US National Cancer Institute, Andrew von Eschenbach, announced an ambitious plan to eliminate cancer by 2015. It was the latest step in a US War on Cancer launched in 1971 by President Nixon. 2015 came and went without von Eschenbach's vision coming to pass. And by 2016, the rhetoric had shifted, as President Obama launched a less time-constrained "national 'Moonshot' initiative to eliminate cancer as we know it."

Cancer is a vicious killer and destroyer of human lives. In 2018, there were an estimated 9.6 million deaths from cancer around the world, with around 17 million new cases of the disease appearing annually. It's not the top cause of death—cardiovascular disease still holds pole position here—but it is one of the hardest to come to terms with. And as a result, it's not surprising that eliminating cancer is a top priority for scientists, politicians, medical professionals, and many others.

Yet, while detection and treatment technologies have advanced considerably in recent years, the global burden of cancer on society isn't decreasing nearly as fast as many people once believed it would.

If the War on Cancer has taught us anything, it's that our biology is far more complex than we originally thought, and that easy fixes to complex problems are often elusive. Driven by the hubris of believing that, if only we understand how cancer works, we can fix it, we've learned the hard way that we don't have the ability to find solutions to every problem. Just because we are smart and can imagine a future that's different from the present, doesn't mean that it's automatically within our grasp to achieve.

Of course, vision and ambition are important in building the future. We may only see through a glass darkly in the present, but it's our belief that we can build a better future *despite this* that often keeps us going. Our excessive self-confidence is what inspires us to achieve more than sometimes seems possible, whether we're focused on curing cancer, eliminating poverty, or establishing communities on Mars. Our hubris pushes us forward to take small but important steps toward building the future we want, even if the ultimate vision remains elusive.

And yet, because of the unpredictable connections between cause and effect, there's a danger that we become too wrapped up in hubris, and allow our pride to blind us to the realities of what's possible and what is not.

Hubris, if we're not careful, leads to false hope. It tempts us to promise what isn't within our power to deliver, and it encourages belief in the absence of evidence. And the danger is that, when hubris fails to deliver, hope too easily turns to doubt and despondency.

The future is slippery. It can be elusive and unpredictable, and it often doesn't align with what we plan. Without care, hubris becomes a bait-and-switch where we're tempted to make a material and emotional investment that is never going to pay off.

There's also the risk of unintended consequences, as hubris blinds us to thinking critically about what might go wrong as we strive to build the future we desire. It's easy to believe that, with enough time and investment, science and technology can transform any imagined future into reality. It's inspiring, motivating, and exhilarating to be at the cutting edge of ideas that we know with absolute certainty are going to solve the world's problems. Yet such naive myopia too often ends up causing a trail of destruction as the hidden consequences of such hubris reveal themselves. And nowhere is this more apparent than when hubris tips over into delusion.

41

———

DELUSION

In April 2018, Mike Hughes launched himself over five hundred meters into the air while strapped to a homemade rocket. His mission: to prove that the world is flat.

Hughes had a very clear and particular vision of the future. It was one where he debunked over two thousand years of knowledge, and demonstrated once and for all that the world is indeed flat. Sadly, no amount of wishful thinking is going to change the reality that he was wrong. But you have to admire his dedication, even though it was eventually his undoing. Sadly, Mike was killed in a failed rocket launch on February 22, 2020.

Delusion is a particularly intriguing aspect of the human psyche. It has a lot in common with hubris in that it enables us to believe fervently in a future that isn't overly concerned with reality. It draws on our capacity for imagination and creativity, and our ability to use our intellect to fill voids in our understanding with whatever best fits our dreams. And, while it's easy to criticize the specks of delusion caught in the eyes of others, it's often much harder to see the log sticking out of our own—because no matter how reasonable we think we are, each of us has our own delusions about the future.

For most of us, these delusions don't rise to the level of those entertained by flat-earthers. And yet we're biologically predisposed to create mental models of the world we live in, and the future we're building, that don't always match reality. There are even indications that the smarter we are, the better we are at justifying our beliefs *in spite* of evidence to the contrary.

Of course, the line between inspired imagination and delusion is preciously thin. Our power to change what's coming down the pike arises from our ability to imagine futures that are different from the present. Yet, when our imagination becomes separated from reality, it slips comfortably into the realm of fantasy.

Ironically, ungrounded beliefs can still have a profound impact on our future. Believing fervently that you can fly, or that you're invincible, is likely to lead rather rapidly to a future with a "you-shaped hole" in it. More insidiously, a belief that vaccines are highly dangerous, or that homeopathic remedies can cure cancer, or that climate change has nothing to do with human activities, can all deeply impact the future we inherit. What we learn from delusion is that it's not always what we *know* that leads to actions that influence the future, but what we *think* we know, or what we *perceive* to be true, that matters.

42

PERCEPTION

There's a saying among people who study risk that perception is everything.

This is, of course, strictly speaking, not true. No matter how much you fear flying, it isn't going to affect the likelihood of a crash (unless you're the pilot, of course). Likewise, phobias tend to reflect a disconnect between a fear of something—spiders for instance, or open spaces—and the likelihood of them causing physical harm.

And yet, the decisions we make in the present are ultimately based on how we imagine the future to be, and this in turn is colored by how our wonderfully complex, and most definitely flawed brains, perceive the present.

We are constantly creating maps in our minds that describe where we are now, and where we could be a few steps into the future. These maps are built up from signals and inputs from the world around us—what we see, what we hear, and what we perceive to be real and true. In our heads, these maps draw on what we know. And where there are gaps in our knowledge and understanding, our imagination and creativity fill them in.

As a result, we instinctively create a vision in our mind of what our world may look like in the future. As we do so, we make calculated guesses as to whether this future is likely to cause us pain or pleasure, and we take action to avoid the one while embracing the other. In this way, our perceptions color how we think about and move toward the future. And sometimes, they lead to us getting things wrong.

Perception can be a life-saver as it alerts us to impending dangers, but it can also fool us into thinking things are risky that are, in fact, not. Perception affects how we respond to situations that look dangerous. It's what leads to us favoring products that we believe are safer, irrespective of the evidence. And it underpins who we choose to trust and who we decide not to.

Because of this, our ability to design and build a better future depends on how able we are to recognize when our perceptions diverge from reality, and to correct our course when they do. This is possible, but it takes awareness and discipline. And, just to make things more difficult, we live in a world where it seems increasingly easy to be deceived into believing that our perceptions of reality are, in fact, valid— even when they're not.

43

DECEPTION

How we perceive the future, and how we envision the various pathways toward it, depends to a large extent on the information we have at our disposal in the present. We develop our mental models of the future based on what we've experienced and what we've witnessed, along with what we've been told, what we trust, and what we believe to be true.

But what if the reality we're building our future aspirations around is, in fact, a fake?

The art of deception has a long, if not necessarily illustrious, history. Ever since we've been able to make use of our imagination and creativity, we've been weaving alternative visions of reality that are intended to fool others. Con artists, marketers, politicians, sociopaths—they all depend on playing to our limitations as they persuade us to buy into a future that serves their purposes, but not necessarily ours.

These are the masters of deception, the "fake future" artists, and they are good at their trade. Thankfully, most of us have a finely tuned antenna for spotting deceptions. However, technology is beginning to challenge this.

Over the past couple of years, there's been growing concern over so-called "deepfakes." These are manufactured videos of people saying and doing things that are so realistic they're all but indistinguishable from the real thing—apart from the content.

Despite the increasing use of photorealistic computer-generated imagery in movies, we tend to trust video from sources like news feeds, bodycams, or even smartphones. We inherently accept that these inform us of the truth of the present, and the pathways we need to take to navigate to a better tomorrow. But if these trusted sources of information become corrupt, where does this leave our visions of the future?

Of course, a video of a politician behaving out of character is a dead giveaway that there's probably a faker behind it. But how about deep fakes that so incense or enamor us that our "fake-o-meter" simply doesn't kick in? Just how vulnerable are we to having our perceptions of the future clouded by fake videos of protesters, for instance, or police brutality, or even terrorist activity? Where our minds are primed to believe we're heading for a dystopian future that's going to take radical action to avoid, how susceptible do we become to deepfakers who are intent on nudging our vision of the future toward one where they are the ultimate beneficiaries?

The hope is that we learn to inoculate ourselves against such fakes, including using techniques like seeking out multiple sources of information before jumping to convenient conclusions. Yet even with these precautions, deceptions are a growing part of a broader landscape of challenges that threaten our relationship with the future we aspire to.

44

THREAT

Late in 2018, Sal Parsa and Joel Simonoff launched a new service to help parents find the right babysitter. By using AI to profile candidates based on their social media history, their app Predictim rated potential babysitters according to how trustworthy and reliable they were deemed to be.

Predictim seemed, on the surface, a great idea. What parent doesn't want to do everything in their power to protect their children against potential threats? However, as word spread about the service, there were mounting concerns that babysitters were being unfairly scrutinized and assessed—and by a machine at that. Following an article in the *Washington Post* that raised substantial concerns around the validity and appropriateness of the app's risk ratings, Predictim was eventually pulled from the market.

What fascinates me about the Predictim story is the way it reveals just how tortuous the path between the present and the future can be—especially when something of value is threatened. The whole premise of Predictim was to protect young children through ensuring that their caregivers were not going to impede their journey to a bright future. And yet in doing so, it threatened the dignity and livelihood of these selfsame caregivers. And it's this that in part led to the service's demise.

Threats to what we value, it turns out, have a powerful impact on how our future unfolds. We're used to thinking of such threats in terms of risks to our health, or to the environment we live in. But the decisions we make are influenced by threats to a much longer list of things that we value. These include our sense of self-worth and social acceptance, our dignity, social justice, ethical behavior, the ability to control our own lives, and many more. Threats to things of value like this are rarely quantifiable, and as a result they are easily overlooked. But it's threats like these that litter the landscape between where we currently stand, and the future we aspire to. And we ignore them at our peril.

There's a name for these types of threats: "orphan risks"— so called because they're all too easily overlooked or "orphaned." These are risks that lie between the present and the future, and that are visible if you look for them, but which are too often ignored because they're not deemed important enough to pay attention to. And yet, they inevitably end up being the ones that trip us up—just as Predictim's lack of awareness of the social risks its technology created contributed to the company ultimately failing.

Orphan risks can be as diverse as the objects, beliefs, rights, and aspirations we hold to be important. And many are ingrained in our sense of who we are, and who we seek to become. Sadly, there are few risks more devastating than being denied the future you aspire to, simply because of who you are. These threats to value are risks that we can learn to spot and navigate around—if we work at it. But if we don't, they will continue to lie in wait for us on the road to the future, and, just as Predictim discovered to their cost, they have a habit of blindsiding us when we're least expecting it.

45

———————

BLINDSIDE

"I never saw it coming" is how most stories of blindsides start. Whether it's the breakup of a romantic relationship, a car crash, or an unexpected sporting maneuver, blindsides trip us up on our journey toward the future and, by their very nature, always take us by surprise.

Blindsides come with the territory of living in a complex world. They occur in the liminal space between how we imagine the future playing out and the unknowability of what's going to happen next. And they're an inevitable part of stepping into the unknown as we bravely move toward the future.

Not that any of this makes them less painful.

Blindsides can be deeply personal. But, as we saw with Predictim, they also affect businesses and other organizations. Politicians have known this for as long as there have been politicians—especially the ones who suffer the consequences of misreading the political landscape. And, as many technology companies are discovering, when you introduce novel technologies into a deeply complex and interconnected world, the road between the present and the future becomes increasingly uncertain, with metaphorical bends, potholes, and dead ends turning up when least expected.

Social media companies are currently going through a blindside minefield as they discover that their users have very clear, but not always similar, ideas about privacy and free speech. Food and agricultural corporations are still suffering the aftershocks of being blindsided by what was, to them, an unexpected public backlash against genetically modified foods. And we're just beginning to discover the extent of the potential blindsides that lie along the path between what we can potentially achieve with AI and its socially responsible use.

Because the future is always breaking new ground, there will unfailingly be blindsides on the way. And as technologies become increasingly powerful and complex, the potential for substantial blindsides is only going to increase. Yet this doesn't mean that they're always inevitable. And the secret to avoiding them is buried in the term itself.

Blindsides exist because we lack the ability to perceive what is about to trip us up. But what if we could pry open the metaphorical blinkers shading our eyes from the future, even just a little, and see more clearly what lies in wait as we move toward it? If we were able to develop a greater awareness of what's in front of us as we navigate toward the future, we should be able to develop ways of spotting at least some of the potentially blindsiding events that stand in our way. One way to do this is to open ourselves up to new ideas, and to listen to people with very different notions, beliefs, and perspectives from ours. This may sound simple, trite even, but it's a way of beginning to better see the landscape between where we are and the future we're heading toward that can be surprisingly revealing.

This becomes especially important where the consequences of a blindside are potentially catastrophic. Where there's no turning the clock back on ill-informed decisions or serious miscalculations, we need to collectively get much better at avoiding blindsiding snares along the way to the future. These include dangers such as maiming or killing people with seemingly good technologies, or the creation of capabilities that undermine human dignity and autonomy. And they most definitely include actions that harm the environment which, if left unchecked, have the potential to change the world we live in for the worse.

Yet, while blindsides are unwelcome, they are a result of the greater reality that we live in a world that is dominated by change. And as a consequence, the more adept we become at navigating change, the better equipped we'll be for building the type of future we want.

46

————

CHANGE

Change is the lifeblood of our existence. It governs the planet we live on. It marks the passage of our lives from conception to death. Change is at the heart of innovation and invention, and of learning and growth. It permeates our imagination and creativity. And it is indelibly woven into our deepest hopes and aspirations.

Change is simultaneously a medium we travel through, a force we reckon with, and a doorway to untold opportunities. If we were left to the whims of change, our history as a species would be very different. Over millennia, however, humans have learned how to understand and predict change, and use it to their advantage. And nowhere is this clearer than in the discovery of the mathematics of small differences, better known as calculus.

Truth be told, calculus, for most people, is at best an irrelevance, and at worst, part of their tortuous and seemingly unnecessary rite of passage between grade school and what lies beyond. And yet it's sobering to think that, without calculus, few if any of the innovations we now rely on would have been possible. The cars we drive, the clothes we wear, the food we eat, the electricity that powers our homes, and so on—none of these would be possible without this branch of mathematics.

Calculus gives us the mathematical and scientific tools to work with differences. These may be differences associated with the shapes and contours that define the three-dimensional world we live in. But where these differences involve time, calculus becomes part of a powerful mathematical language of change, and one that enables scientists, engineers, economists, artists, and many others to understand and use the transition from past to future to their advantage.

Perhaps not surprisingly, this is a far-from-perfect language, as our lives are defined by more than mathematics alone. And yet it's one that has enabled humanity to take huge strides in overcoming the uncertainty of being adrift and rudderless in a vast and unending sea of change.

Thankfully, and despite this, not everyone needs an advanced degree in calculus to thrive in the modern world. But when we collectively set out to design pathways toward the future we hope for, this ability to understand, model, and work with change is pivotal to our success.

Using the mathematics and science of change, it becomes possible to get a sense of where observed trends and trajectories are heading, and where they're likely to speed up, slow down, stay steady, or die out. The same science and math help identify impending dangers, as change threatens to take away what we value. They also play a crucial role in building resiliency and agility in the face of change as we work toward building a better future.

Of course, these admittedly powerful ways of understanding and utilizing change have their limitations—the future is, after all, a journey into the unknown, and complexity and uncertainty can confound even the most sophisticated of predictions. And yet, when tempered with humility and guided by our humanity, our technical mastery of change can help set boundaries around what we don't know or cannot predict, and it can reveal pitfalls that may otherwise deeply harm us on our way.

As we stand on the edge of tomorrow and look out toward an uncertain future, this ability is becoming more important than ever as we chart our way across an increasingly turbulent sea of change. And yet, for all the power that our fluency in the language of change endows us with, it's worthless if we don't have a clear eye on where we're heading, and why—especially as the world we live in comes under increasing stress from the demands we make of it.

PART 4

THE EDGE OF TOMORROW

"...there's no single answer that will solve all of our future problems. There's no magic bullet. Instead there are thousands of answers—at least. You can be one of them if you choose to be."

—Octavia Butler

47

RESTRAINT

In 1965—the year I was born—there were 3.3 billion people on Earth. By 2015, this had more than doubled, to 7.5 billion. And at the current rate of change, we'll have exceeded 10 billion by 2065.

This is not only a growing population, it's also one with growing expectations and demands. And it's living off a planet that has only so much to give. The rate at which we are using precious global resources and pushing the boundaries of what the Earth can withstand and bounce back from is, if anything, accelerating. And as it does, we're looking at a prognosis for the future that, at the current rate of change, doesn't look so good.

It's a stark reality of living in a universe governed by change that, for every action in the past, there is an associated reaction in the future. We know this, of course. We've evolved the creativity to imagine a future that's different from the present we inhabit, and we've developed the intellect to predict what that future may look like based on past and present trends. Unfortunately, we've also inherited a stubborn streak that, all too often, blinds us to the type of future we're heading toward.

If we apply the science and mathematics of change to our current trajectory, the outlook looks bleak. We can map out a future where more people are demanding more food, more energy, and more water, from an increasingly stretched planet; a future where the continuous environmental stresses of greenhouse gas emissions and other forms of pollution continue to weaken the planet's resilience; and a future where the confluence of human power, selfishness, and sheer foolishness conspires to destabilize societies around the world.

These are not accurate predictions, of course. But like weather forecasts, while they may miss the mark on specifics, they lay out highly probable scenarios, unless we actively intervene.

The irony, of course, is that we're living through a period in human history where we have an astounding technological and social toolkit at our fingertips for designing and engineering the future. And yet, at times, it seems that, despite this, we're more interested in indulging our predilection for short-termism and self-preservation, to the detriment of billions of people in this generation and those to come.

In our journey into the future, we've developed quite amazing abilities to help craft our way toward a world that is substantially better than the one we currently inhabit. But we're still struggling to learn how to wield them responsibly. And nowhere is this more apparent than in the ways we are blindly pushing past no-go boundaries that have successfully constrained and guided the future of Planet Earth in the past.

48

BOUNDARIES

In 1980, the Cambridge physicist Brian Pippard published a paper describing what he called "experiments in critical behavior and broken symmetry." In it, he explored particular types of transitions between the present and the future that he referred to as "discontinuities"—the blindsides of the physical world.

Pippard was fascinated by transitions between present and future that were abrupt and irreversible—transitions that occur at a tipping point beyond which everything changes and there's no going back, such as the snapping of a branch, or the breaking of a wave.

I probably wouldn't be aware of Pippard's work if I hadn't attended one of his public lectures as a PhD student. In the lecture, he held up a simple model of a vertical ladder consisting of four evenly spaced wooden rungs, held together by two lengths of string. He then asked the audience what would happen if he slowly rotated the bottom rung through one complete horizontal revolution. Naturally, we predicted that Pippard's model would smoothly transform from its conventional ladder-like form into something that looked more like an artist's impression of a strand of DNA—a neat double helix, consisting of two lengths of string held apart by the wooden rungs. And of course, our vision of the future was utterly wrong.

As Pippard twisted his ladder, the lengths of string between two of the rungs suddenly twisted together, destroying any semblance it had to DNA. The result was a tangled mess.

This was an abrupt and irreversible transition—reversing the twist failed to untangle the ladder. But the point at which it occurred—and the point at which some irreversible boundary was crossed—was all but impossible to predict.

Over the intervening years, it's become increasingly common to talk about tipping points—hard-to-predict points of instability in seemingly stable systems—especially in the context of climate change. Just as Pippard's ladder demonstrated, there are concerns that we're in danger of crossing such boundaries that mark a point of no return as we continue to stress the environment. And if we do, we risk disrupting, and even destroying, critical pathways to the type of future we'd like to see.

Pippard's ladder is an example of nonlinear dynamics. It represents the tendency of complex and interconnected systems to undergo rapid and irreversible changes when stressed. And it's a sobering reminder that, even though things may look great in the present, unless we learn how to spot early warnings and stay clear of critical tipping points, we run the risk of, quite literally, crashing our future.

Similar types of behavior are seen wherever complex, interconnected systems exist. And nowhere do they raise more concerns than when they involve the planet on which we live. There are growing fears that, just as Pippard's ladder undergoes a sudden and irreversible transition when pushed too far, we're getting perilously close to overstepping similar critical boundaries here on Earth.

As human activities continue to stress the world around us, scientists are keeping a careful watch on such boundaries, and how far we are from overstepping them. These include the ability of the Earth's climate to resist and absorb the impacts of non-renewable energy use, the diversity of living organisms on the planet and the genetic richness they represent, and the capacity of ocean-based ecosystems to accommodate increasingly acidic conditions.

As we continue to push toward these and many other boundaries, it's not clear how our planetary future will unfold. But from everything we know about complex systems, it's a fair bet that it won't be as conducive to human life as we hope, unless we take action to avoid these points of no return. Yet even without the helping hand of humanity, one of the consequences of living on a dynamic planet is that our future is sometimes determined by cataclysmic events, over which we often have little control.

49

CATACLYSM

On June 30, 1908, a massive explosion struck eastern
Siberia, flattening over seven hundred square miles of
forest, and breaking windows hundreds of miles away. The
explosion was the result of a meteor strike—most likely
occurring as the meteor broke up in the Earth's atmosphere.
It was the largest event of its kind in recorded human
history, and it stands as a monument to just how rapidly
and unpredictably the future can change.

Meteor strikes of this magnitude are exceptionally rare.
Collisions that have the potential to dramatically affect lives
around the globe only occur every few hundred thousand
years or so. They are, however, a reminder that we live in
an uncertain world that is constantly subject to dramatic,
future-changing events. And if you're at the epicenter of one
of these, the results can be devastating.

Natural events keep us on our toes as an evolving species,
and help prevent our hubris from getting the better of us.
They stand as a caution that, no matter how confident we
are in our ability to design the future, there's always the
possibility of a metaphorical roll of the dice that radically
changes the playing field. This may be a massive volcanic

eruption, a catastrophic earthquake, or a deadly pandemic. It could, on the other hand, involve something less obvious and yet equally devastating, such as the loss of major ice sheets, changes in critical ocean currents, or the world being engulfed by massive solar flares.

Fortunately, events like these that have the potential to profoundly affect large populations around the world are rare. Yet if you are personally caught up in even a local incident, it can sever your ties with your anticipated future in ways that can be just as devastating.

This was the case for the inhabitants of the Roman city of Pompeii in 79 AD, as Mount Vesuvius erupted and entombed them in deadly ash flows. Similarly, the Indian Ocean earthquake that occurred on December 26, 2004, utterly changed the future for hundreds of thousands of people as the resulting tsunami swept along coastlines, leaving over two hundred thousand dead in its wake. For these and many others throughout history, their vision of the future was altered in a heartbeat, as the Earth metaphorically shifted itself into a more "comfortable" position.

Yet, fickle and unpredictable as the planet we inhabit is, we have also become masters at bringing about cataclysms of our own making, and shattering futures through our own ineptitude. In December 1984, thousands were killed around the Indian city of Bhopal as toxic gas enveloped them from a local pesticide plant. In 1986, an explosion in reactor four of the Chernobyl nuclear power plant led to the deaths of thousands from radiation exposure, and rendered the entire region uninhabitable for decades. And as we push the Earth's climate to breaking point with greenhouse gas emissions and other forms of environmental pollution, we are further courting catastrophe.

Of course, we could fatalistically accept that we live in a precarious world that's governed by short-sighted individuals and institutions. But why should we? As humans, we have evolved an exquisite ability to envision the future, and to design and engineer it. And while we have a destructive streak that seems intent on leading us to alter the future with no thought to the consequences, we have the capability of overcoming at least some of our limitations. We may never be able to avoid every cataclysmic event. But we do have the capacity to manage those that we have influence over, and to build resiliency against the ones we don't.

It's almost as if this is a rite of passage we need to go through as we mature as a species, part of a natural progression from our biologically-constrained childhood and technologically tumultuous teenage years to a grown-up community that understands its responsibility to the future, and takes it seriously.

As we continue along this progression, the speed with which we mature as a species will be important in determining our fate. If our growing capabilities exceed our collective ability to use them wisely, there is likely to be trouble ahead. Some would argue that we've already hit this point. And not surprisingly, there's already talk in some quarters about abandoning Earth altogether as a result, and beginning afresh elsewhere in the solar system.

50

OUTWARD

In October 1982, Walt Disney World's Epcot Center opened its doors for the first time. Its inaugural visitors encountered a futuristic-looking geodesic sphere as they entered the center—a feature that continues to welcome people to this day.

Epcot's "Spaceship Earth" is an iconic exhibit that takes visitors through the history of communication technologies and encourages them to design the sort of future they want to live in. Its inspiration, though, comes from a long history of imagining the Earth as a self-contained ship, hurtling through space.

This idea was captured in economist and writer Barbara Ward's 1966 book *Spaceship Earth*. Ward imagined us inhabiting a planet hurtling toward an uncertain future, while grappling with limited resources and complex social and political challenges. Just a few years later, the architect and inventor of the geodesic dome, Buckminster Fuller, pushed the idea further in his more popular and ultimately more influential book *Operating Manual for Spaceship Earth*. In it, he crystallized the image of Earth as a planetary-scale spaceship wending its way through the cosmos.

Buckminster Fuller's *Operating Manual* takes an expansive look at humanity's journey into the future. "We are all astronauts" on a journey through space, he wrote, at a time when real-world astronauts were quite literally breaking away from Earth's grasp. In the manual, he took a hard look at what it'll take to keep this spaceship functioning smoothly. But he also sowed the seeds of thinking about what lies beyond our current limitations.

Buckminster Fuller, and many others since, have warned of the need to plan carefully for the future if we're to avoid overstressing the Earth's limited resources. The result has been a vibrant if perpetually-frustrated environmental movement, as conflicting visions of the future, and how to get there, play out. But the idea of us occupying a ship hurtling through space has inspired another strand of future-thinking that is gaining ground in some circles: What if we're in danger of making such a mess of things on "Spaceship Earth" that we need to start thinking outside the box, and find another ship?

This is a "plan B" that, in at least one version, involves quite literally jumping ship to somewhere like Mars, or one of the less harsh moons of Jupiter or Saturn. This may feel like science fiction, and to be quite frank, this is how it's likely to remain for the next several decades at the very least. Yet over the past few years, two pioneers of private space exploration have taken it seriously enough to sink hundreds of millions of dollars into a plan B for the future.

Amazon founder Jeff Bezos is so convinced of the importance of space exploration to our future that he's invested a large chunk of his personal fortune into his aerospace company Blue Origin. Bezos believes that, as Earth's resources become increasingly strained, our only hope is to start mining planetary bodies for precious supplies. In his mind, our future is inescapably tied up in space exploration as a means of continuing to support the existing "Spaceship Earth."

In contrast, Elon Musk's plan B is to escape Spaceship Earth altogether. He's so convinced that the Earth's future looks bleak that he's planning to establish communities on Mars as an escape route, and is building the capacity to do so through his company SpaceX.

Both of these visions of the future have their issues— especially when you consider that most of the problems we currently face have their roots in humanity's inability to act responsibly toward the future. Sadly, there's no evidence that we've grown out of this phase yet. But even if we had, there are near-insurmountable challenges to establishing a sustainable human presence in space, not least because it's a "fail-dead" environment (unlike much of Earth), where even the smallest error can lead to a future that is extremely short indeed.

And yet as Bezos, Musk, and others are actively demonstrating, space has a deep fascination for many of us as we think about the future. The more we learn about the solar system and beyond, the easier it becomes to imagine a future where we embrace this new frontier—aided and abetted by a rich history of space-inspired science fiction. It's a vision that is inspiring and motivating to many people as they look outward to the stars, much as Anders's *Earthrise* inspired a generation in 1968 as it provided a vision of Earth from space.

As we imagine and plan for this outward-facing future, however, a niggling question keeps recurring: are we alone in the universe? Or are there new lifeforms to be discovered, and new intelligences to be encountered? And if there are, how will this affect our own unfolding future?

51

LIFE

The cosmologist, writer, and science popularizer Carl Sagan is famously credited with observing that "The universe is a pretty big place. If it's just us, it seems like an awful waste of space."

Sagan was right, of course, if life, and intelligent life in particular, are the inevitable expression of the driving forces behind the universe's progression from past to future. While we still don't know if we're alone or not, the more we discover about our neighboring planets and moons—including the existence of Earth-like planets in other solar systems—the more likely it becomes that we'll find evidence of life elsewhere.

When we do, though, it's not clear how this will disrupt our visions of the future, and the pathways we're building to get there.

Our visions of extraterrestrial life, and how its discovery might potentially impact us, are ultimately constrained by what we know. Just a few decades ago, we knew next to nothing about the possibility of alien life existing, and so we filled the void in our heads with fantastical ideas. These fantasies colored our visions of the future, but not to the extent that they substantially influenced our actions in the present—at least, not for most people.

Then, scientists began to discover that some terrestrial organisms can thrive under seemingly impossible conditions. These are the "extremophiles"—organisms that can survive around hot thermal vents, in lakes buried far under the Antarctic ice, or even deep in the Earth's crust. Researchers then started to find evidence of astronomical objects that could potentially harbor their own forms of such extremophiles. We began to detect traces of chemicals that could be the precursors of life, far beyond the limits of the Earth. And we started to find planet after planet in the galaxy that could, in principle, support the emergence of life not too dissimilar from that found here.

As a result, in just a few short years, the chances of us discovering life elsewhere have shot from being highly unlikely to looking increasingly probable. And with our changing perspective, our ideas about the future, and how it in turn defines and guides us in the present, have begun to shift.

Sadly, the chances of us discovering intelligent life that didn't originate on Earth are remote—not because it's not there, but because, in the vastness of the universe, the likelihood of our paths crossing is infinitesimally small. And yet, as scientific efforts to detect evidence of extraterrestrial life in the solar system ramp up, there's a growing possibility that, in the coming decades, we'll discover that we're not the only place in the universe where life has existed.

Whether this will shatter our preconceptions of the future and cause us to rebuild them in new and intriguing ways, or lead to us simply hunkering down in the comfortable but increasingly erroneous stories we tell ourselves about how unique and special we are, remains to be seen. But as we probe our planetary neighborhood and look to distant galaxies and beyond, we're likely to find something that, one way or another, profoundly challenges how we think about the future.

Yet as we embark on this grand adventure, there is another search for "alien" life that is going on, and one that could impact our vision of the future just as profoundly as if we discovered the existence of extraterrestrial organisms. And this is the quest to create, through our own ingenuity, the first completely artificial life forms here on Earth.

52

RE-CREATION

In May 2019, a team of scientists announced they had created a "designer" bacterium with a completely synthetic genome. The organism was based on the common *E. coli* bacteria which inhabits our guts, and which is a workhorse of bioengineering laboratories the world over. What set this bacterium apart, however, was that its DNA—its biological code—was designed and engineered from scratch, using common laboratory chemicals.

Ever since the discovery of DNA, scientists have been intrigued by whether it's possible to use this genetic code to redesign living organisms. In principle, such a "re-creation" of life is possible—DNA is, after all, simply a collection of molecules arranged in a specific way. But it's only since the advent of cheap and fast DNA sequencing and synthesis, together with powerful computational capabilities, that we've come close to being able to achieve this.

With recent advances in technology, we've progressed to the point where it's possible to create highly complex genetic sequences by design, and we're getting better at inserting these sequences into living organisms. We are, in effect, getting remarkably close to designing new organisms as easily as we design new apps.

When we reach this point, anyone with access to the necessary technology will be able to design the future in ways that will make our previous efforts look like child's play. We will, for all intents and purposes, be creating "alien" life on Earth.

This possibility alone is enough to shake up our visions of the future, as what was once relegated to the realm of fantasy begins to look increasingly plausible. And as we begin to flex our DNA-manipulating imagination, we're likely to get increasingly good at making these dreams of lab-grown life come true.

There are, however, substantial hurdles we still need to overcome on the road to this synthetic future. We can currently design increasingly long sequences of DNA using computers, and we can painstakingly reconstruct these in living organisms. But we still don't have a good grasp of how to interpret the code of DNA. And we have an even poorer understanding of how our heritage, our history, and our everyday environmental interactions influence our "epigenome"—the genetic overlay that constantly affects how our genetic code is expressed.

To overcome these limitations, scientists are increasingly turning to machines to help them design and engineer DNA. In effect, another creation of ours—artificial intelligence—

is being used to help us create and manipulate biological life. This is a confluence of capabilities that is making the future of home-grown "alien" life ever-more likely, but increasingly difficult to predict.

As we get closer to creating new life, whether this is biological in origin or embedded in intelligent machines, our visions of the future are going to have to adjust accordingly. This is not going to be an easy transition, as we rethink our assumptions and expectations based on billions of years of evolution. And one of the challenges we'll face is how our emerging capacity to create a designer future affects how we understand a fundamental aspect of our own lives: what it means to be human.

53

HUMANITY

For most of us, our vision of the future is uniquely human. It's driven by the skills, attributes, and perspectives we've evolved to be capable of, and it's intimately connected to who we think we are, and who we aspire to be. And yet, what it means to be human, or to be imbued with "personhood," is surprisingly elusive. In the words of author John Green, "the deeper I dig, the harder it becomes to understand what makes people, people."

Biologically, we're merely a by-product of a constantly evolving environment. And genetically, we're not that different from many other organisms. But of course, it's the differences, deceptively small as they may seem, that help define us as a unique species.

These differences have led to us emerging as animals with an exquisite sense of the future. More than any other organism living on Earth, we are capable of imagining a future that's different from the past, and actively charting a way toward it.

Humans are, in a very real sense, architects of the future. We live in a present that is dominated by our perception of the future and what it might hold. Our every action is determined by what's coming down the pike, and how we can ensure that it's good for us—often, it has to be said, to the detriment of other people and other organisms. Despite our future-oriented ambitions, we're also short-sighted.

Of course, we're not the only evolved organisms that can anticipate and respond to the future. But we've taken it to a level beyond anything seen anywhere else. Through our intellect, our creativity, and our ability to innovate, we are crafting new technologies, new societies, and new worlds. We're on the cusp of designing new organisms, even brand-new forms of biology. And if we crack artificial intelligence, we could be heading toward designing a future in which we are, for all of our capabilities, redundant.

This in-built ability to envision the future, combined with a compelling desire to change it, is part and parcel of what it means to be human. And yet it's only part of the story. Our humanness extends to less tangible qualities that include how we feel and how we behave toward others. And sadly, it includes a tendency to exclude from the future that we're building those we consider to be in some way less valid, less entitled, even less "human" than us.

Up to now in this exploration of the future, I've been rather loose with the term "we." I've implicitly assumed that there's a homogeneous "we," where "we're" building a future that "we" will all benefit from as "we" work together to make it so. And yet, the reality is that the "we" of humanity encompasses a diverse collection of individuals who all have their own ideas of what the future should look like. And the danger is that, as we come into conflict with those who threaten our particular vision of the future, we delegitimize their claim by undermining the very validity of their humanity.

This is a particularly ugly and insidious part of being human. It draws a tight circle around our conception of "we" that conveniently excludes those who don't look and think like us, or don't share our views and our visions, or who otherwise threaten what we value—whether that's our worldview, or our greed for a future that's just about us and our desires.

This is a deeply selfish and destructive approach to building the future. Instead, we should be striving for this "we" to be as big and inclusive as possible. And we should be collectively building a future where everyone has the right to thrive, as long as in doing so, they don't deprive others of this selfsame right.

Here, I deeply appreciate John Green's perspective when he says, "I believe that when we acknowledge each others' consciousness and complexity we lead better lives, and feel less alone in our grief and in our joy...I believe that we're human because we believe in each other's humanness and because we can listen and we can work together to alleviate each other's suffering. And in that sense I guess that being human is both something that we are and something that we must always aspire to be."

This is a vision that inspires us to build a future that's inclusive, that puts others first, and that is designed for the benefit of the many, not just the few. It's one that opens the door to understanding what it means to be human and to have "personhood" long after we've outgrown the constraints of our biological heritage. And it gets us thinking about the *why* of future-building, as well as the *how*.

54

MEANING

Imagine you're sitting on a park bench and a dog drops a ball in your lap, then sits back, its tail wagging, as it looks expectantly at you. Do you throw the ball for the dog, or just ignore it?

Whichever you do, you have temporary control over how a small slice of the future plays out. In one version, the dog joyfully chases after the thrown ball. In the other, it slowly realizes that it's chosen the wrong human.

What is it that determines which of these futures you create? Where's the "why"—the meaning—in the decision you make to throw the ball, or to deny the dog its desires?

It may be that you throw the ball because seeing the dog enjoying itself gives you pleasure. On the other hand, you may ignore the dog because to stoop to its antics would, in your eyes, demean you. Or it may simply be that you're annoyed by its unwanted attention. In each case, some part of your brain will have weighed up the consequences of your actions, and made a value judgement about which future to bring about.

Light-hearted as this scenario is, it gives a sense of how meaning is linked to the accrual of value. Committing to an action often has meaning to us because it leads to a future that has greater or equal value when compared to the past. This "value" can be as seemingly simple as doing what is necessary to stay alive and healthy. But we are far from simple creatures, and as a result, the "value" that's associated with meaning can take on many different forms—including the pleasure of watching a dog run joyfully after a ball.

From a purely biological perspective, meaning arises from decisions and actions that trigger our future-looking reward circuits. We eat, exercise, seek the company of others, or indulge in various habits, because they make us feel good. Yet we also have the capacity to derive meaning from experiences that seems to transcend our mere biology, from cherished beliefs to showing and receiving love and affection, watching a spectacular sunset, contemplating a striking painting, listening to music that seems to reach into our very soul, or simply savoring a good cup of tea. And in many cases, this meaning is captured and conveyed to others through the stories we tell.

In these stories, we map out pathways toward futures that have greater meaning for us and those around us, while erecting guardrails to help avoid those that lead to a loss of meaning. And so, as the dog drops the ball in our lap, we might tell ourselves a story about how throwing it will create a future where, because the dog is happy, so are we. As we tell ourselves the story—albeit subconsciously, most likely—we attach meaning to the event and the outcome, and in doing so we trigger a cascade of biological responses that confer additional meaning by increasing our health and well-being.

But there's a problem with this model. In a world of nearly eight billion people, one person's pleasure too easily becomes another's pain. It's impossible to optimize personal meaning for everyone—it's in part why we so readily demean and discount the meaning of others, lest it threatens to deprive us of ours. Yet in designing and constructing our collective future, we cannot all have what we want.

Thankfully, our sense of meaning transcends ourselves and encompasses the happiness and well-being of others. We see echoes of this in the example with the dog, as the ball-thrower experiences pleasure in the animal's joy. We instinctively derive pleasure from the meaning that others experience, and the meaning we can gift to them. In fact, this ability to provide meaning for others is perhaps one of the more important defining traits of our humanity.

Yet, as we collectively envision and strive to build a future with meaning, there will inevitably be clashes and tensions. And over time, we've developed a deep sense of right and wrong to help navigate these.

55

MORALITY

Wouldn't it be nice to think that, if only we tried hard enough, we could create a utopian future where everyone has what they want? Sadly, the laws of time, motion, and human idiosyncrasy mean we're committed to a future where someone, somewhere, is not going to be happy.

Just as we can't be everything to everyone in our personal relationships, we cannot design and build a future that fits everyone's idea of what it should be like. One of the ways we collectively cope with this is to develop a sense of norms and expectations around what a shared "minimum viable product" of a future might look like. These are often self-defined rules that establish what we think of as "good" or "right," versus "bad" or "wrong."

Naturally, because we're a hot mess of social instincts and individual desires, we struggle to agree on a basic set of morals that collectively guide us in designing and crafting the future. We even disagree on the very essence of what morals are in some cases. And yet despite this, there are widely agreed-on morals-based design principles that help guide how we think about the future, and our obligations to it.

For instance, we tend to collectively share a sense that killing or maiming others is probably bad, and that children are precious. And we have a tendency to believe it's important to be kind and caring, and to help others less fortunate than ourselves—as long as it doesn't cost us too much in return.

Most of us also have a deeply ingrained sense of justice. Sadly, this more often than not kicks in when it's someone else who seems to be getting something we think they don't deserve at our expense. But as a species, we have a surprising capacity to be moved and motivated by the injustices we see others experiencing.

These tendencies coalesce into moral frameworks that help guide our decisions around future-building. And as they do, they help add meaning to our actions and aspirations.

Our individual and collective morals act as a set of principles to help guide us as we translate our imagined futures into reality. And they help us feel okay (or at least virtuous) if the future we end up with isn't quite the one we hoped for. They also set the ground rules for punishing those who don't play along, and marginalizing and penalizing those we fear have a vision of the future that is dangerously out of step with the majority's. Sadly, our morality isn't always what it should be.

But our morals merely establish our principles for future-building. When it comes to converting these into actions, we need to turn to ethics.

56

ETHICS

In 2015, the entrepreneur Elizabeth Holmes was named by *Forbes* as the youngest and wealthiest billionaire in America. By 2016, it was clear that her company, Theranos, was founded more on spin than science, and *Fortune* declared her one of the world's most disappointing leaders. As of this writing, she is scheduled to stand trial in 2020 on charges of conspiracy to commit fraud.

The story of Holmes and Theranos is one of ethical ambivalence in the face of a compelling vision of the future. Holmes's vision was one of diagnostic blood tests that required exceptionally small blood samples, and were inexpensive, automated, and readily available through retail stores. The technology was based on research she carried out while an undergrad at Stanford University, and was so compelling to her that she ended up dropping out and founding a startup in Palo Alto to exploit it.

Based on her vision of the future, Holmes was driven to create a world-dominating tech company, and in hindsight, she didn't worry too much about how she achieved this. Sadly, Theranos's technology ended up falling far short of its promise, and a growing chasm emerged between what

Elizabeth Holmes and her company were claiming and what they were able to deliver. What is quite remarkable, though, is that, until everything began to unravel, Holmes was able to persuade so many people to join her on this journey.

If morals ground our personal and collective principles defining right and wrong, ethics are the social application of these. Ethics are norms of behavior that define what we consider right and proper for people to do when others are involved, as opposed to what's inappropriate. And because most ethical frameworks and principles are based on the consequences of actions or decisions, they deeply affect how we work together to build the future.

Because future-building requires compromise, ethics provides us with a framework within which we can negotiate shared futures—at least, they do when they are applied effectively. Unfortunately, we all have a tendency to bend the rules to get what we want, and ethics are not immune. Except that, the more we deviate from ethical norms—or con people into believing we're ethical when we're not—the harsher the backlash becomes when we're found out.

This is something that technology businesses in particular are discovering to their cost. Elizabeth Holmes's Theranos stands out because the company was built on a web of deception as investors were told the tech worked at the promised price point, and that the company was a good bet—none of which, it turns out, was true. Yet many other enterprises are beginning to lose the trust of investors, employees, and consumers as their behavior is being seen as unethical—especially where it threatens to undermine the types of futures others are striving to build.

Of course, ethical frameworks are only effective when there is broad agreement on how they are crafted and used. Where they become an impersonal checklist or a near-meaningless set of rules and regulations, they become a paper exercise that has only limited power to ensure that our collective future-building ultimately benefits humanity.

For us to successfully work together toward building a shared future, we need to collectively buy into the ethical frameworks and processes that guide our actions. The ethics of what is considered socially right and wrong need to have a clear connection to our personal ideas of good and bad. More than this, we need a deep sense of what is good for others as well as ourselves as we build the future. And this means developing and exercising a strong sense of empathy.

57

EMPATHY

Scott Warren—a geography teacher in Arizona—ventured into the heat of the Sonoran desert in the summer of 2018 to provide humanitarian aid to migrants crossing the border. It's a crossing that individuals and families from Mexico and Central America regularly make in the hope of building a better future. Yet with daytime temperatures reaching 120 degrees Fahrenheit, and water all but nonexistent, it's a journey into a hoped-for future that is fraught with danger. And every year, dozens of people die in the attempt.

Moved by the plight of these migrants, Warren and others were part of a movement that strives to prevent their unnecessary suffering and death. They were driven by a deep sense of identity and humanity, as they pursued a way to alleviate suffering. Unfortunately, however, what they were doing was illegal. As a result, Warren was arrested by Border Patrol agents for providing two migrants with "food, water, clean clothes and beds." Thankfully, even though his case went to trial, Warren was found not guilty. But he could have been imprisoned for up to twenty years for his actions.

Warren's case is far from unique when it comes to strangers risking their own future for someone else's. Along with our ability to imagine our future and chart a way toward it, we also have a remarkable ability to put ourselves in the shoes of others, and to help them build their own future—even when it means putting ours on the line.

This ability to empathize with others, to viscerally feel what it is like to be in their position, and to be inspired to support them as a result, is deeply ingrained in our psyche. And it's an ability to see and feel the path between past, present, and future from someone else's perspective that makes perfect sense at a social level.

Where we have no option but to build a collective future together, empathy provides the impetus to work willingly toward a common goal. Ethics provides us with social rules for living and working together, but it's empathy, more often than not, that provides us with the motivation to do so.

Empathy forms within us a deeply-embedded instinct for group survival, as it enables people to work together toward a greater good. It allows us to see the future through someone else's eyes, and to share in their pain and joy as they contemplate their journey toward it—so much so that we derive satisfaction from helping them on their way. And, as we empathize with others, it also helps us to find meaning in the co-creation of a shared future.

Yet empathy, ethics, and morality are not sufficient on their own if we're to build a future where as many people as possible find meaning and joy. For this, we need to get more specific about what is expected of people and communities on our collective journey. And this means coming to grips with our individual and social responsibility for what lies ahead of us.

58

RESPONSIBILITY

In November 2018, the Chinese scientist He Jiankui announced the birth of twin girls. What made their birth remarkable, and what subsequently attracted the ire of researchers around the world, was that He claimed to have intentionally edited their DNA at the point of conception.

He Jiankui's work was widely denounced as irresponsible, as it broke ranks with established ethical and societal norms and practices. Yet it raised a question that's becoming increasingly relevant to emerging advances in science and technology: what does it mean to be "responsible" when you have the power to change the future?

In He's case, the genetic edits he used allegedly provided the twins with enhanced immunity against HIV, while demonstrating the ability to reengineer the human genome before birth. He did this without his work being sanctioned by the international scientific community, and without fully understanding the outcomes of his actions. At the same time, he opened the door to others following in his footsteps, despite the profound ethical concerns around human genome editing.

Gene editing is just one of many fronts on which scientists and technologists are experimenting with abilities that could profoundly alter the course of the future. Similar advances are being made in artificial intelligence, neuroscience, robotics, and many others. The people working at these cutting edges of science and technology are, in a very real sense, future-makers. They are developing tools that could irreversibly alter the future for billions of people. And in the process, they are opening the way to some futures and closing it to others.

To many, this is part of an exhilarating roller-coaster ride into the future, where the possibilities of what we might achieve far outweigh the risks. And yet, unless this power is wielded responsibly, there is a growing likelihood that, in our blind ambition to build a technologically advanced future, we will overstep the mark and be left with a social and environmental train wreck.

Because of this, there's growing interest around the world in what it means to innovate responsibly, and how we can build a future together where we all have the chance to thrive and follow our dreams. Yet our collective responsibility extends far beyond the capabilities of technology innovation alone. As we intentionally alter the pathway between past and future, whether through science and technology, social movements, shifting norms and expectations, or by other means, we need to take individual and collective responsibility for the potential outcomes of our actions. And this is especially so when the resulting future isn't to everyone's liking, or has the potential to benefit some at the expense of others.

This collective responsibility to our shared future becomes deeply relevant as we look back on the social and environmental impacts of innovation over the past two hundred years. With hindsight, it seems that we've become so enamored with our future-building capabilities that we've taken our eyes off the ball of unintended consequences. And as a result, while overall quality of life is still on the rise around the world, our capacity to ensure this into the future is looking shaky. Social and political trends and movements are disrupting our ability to co-create a vibrant global future. Increasingly powerful technologies are being used without us fully understanding the potential downsides. And as we begin to extend beyond the sustaining constraints of planetary boundaries, we are increasingly in danger of sacrificing our long-term future for short-term gain.

Without a doubt, our collective future—that tantalizing object we hold in our mind's eye—could be profoundly better than the present. If we're to make it a reality, we need to learn how to future-build responsibly, drawing on every ounce of our creativity, inventiveness, and empathy. More than this, though, as creators of the future, we need to learn how to look after and care for what we construct, so that generations to come can benefit from it and add to it.

59

STEWARDSHIP

I don't often tear up when attending conferences. Of course, viewing tear-jerker movies on the flights there and back is another matter entirely—I remember John Green's *The Fault in Our Stars* having me in floods on a flight when it first came out. But conferences? Not so much. Yet I couldn't help myself while attending a panel discussion on the last morning of the 2011 World Economic Forum Annual Meeting in Davos.

The panel was chaired by the Forum's founder Klaus Schwab, and included the then-French Minister of Economy, Finance and Industry, Christine Lagarde, together with three remarkable young people: Nick Vujicic, Raquel Helen Silva, and Daniel Joshua Cullum. The topic of the session was inspiration. But what it indelibly left me with was a deeper appreciation of what it truly means to be a steward of the future.

As the panelists spoke of their experiences and perspectives, they painted a picture of the future where each and every one of us has the responsibility to inspire others and add value to their lives. But what moved me was that these weren't just platitudes. Each member of the panel had grappled with hardship in some way, and had come through with a greater sense of what they had to give to others. And I have to confess that what tipped me over the edge was how profoundly the selflessness of the three young people on the panel affected their audience—including Schwab and Lagarde.

I still think about that panel as I consider the future and our collective responsibility to it. None of us has to be a steward of the future. There's no universal law that says we can't be selfish—although there are always consequences to our actions. And yet, whether through some random quirk of the universe, or by some inscrutable divine intervention, we've been given the gift of being able, not only to imagine and design the future, but to take joy from building a future that values, nurtures, and celebrates others. We have a profound ability to give to others around us, and even to generations to come. And as a result, we are, by design, stewards of the future.

What remains unclear, however, is whether we have what it takes to be *good* stewards.

Sometimes, the evidence is dispiriting. Selfishness and short-term gains all too easily seem to supplant our ability to envisage, care for, and nurture distant futures. And as our ability to alter the future becomes more powerful, there is a worryingly large gap emerging between what we can do, and what we should be doing. This gap has already led to widespread pollution and global warming, and is exacerbating social evils in the name of progress around the world. And it's a gap that too often seems to defy our best efforts to invest in the future and its well-being.

And yet, looking back to that 2011 panel in a snowy Alpine town, I'm filled with hope. If the upcoming generation can imagine a different pathway forward, and inspire a room of hardened global movers and shakers to tear up as they do, there's surely still a chance for us all to learn how to be good stewards of the future as, together, we learn to build something that is better than the present.

60

FUTURERISE

On December 24, 1968, William Anders's *Earthrise* galvanized a generation into thinking differently about the future. It made us realize how precious and fragile this planet we live on is, and what a profound responsibility we have to its future and the futures of those who inherit it. And yet, half a century on, we face more challenges than ever as we struggle to reconcile our actions with their future consequences. At the same time, we have a greater ability than ever before to not only envision a better future, but to work together to make it so.

As we grapple with this growing tension, what will be the equivalent of this generation's *Earthrise*? What vision will inspire us to leave the ruts of narrow-mindedness and selfish short-termism, and strive together to build a better future? What imagined "object" will show us the way to becoming what we can be, rather than being resigned to what we assume is inevitable?

It may be images associated with disaster that motivate us to take action: nuclear war, a devastating pandemic, or the collapse of social order. Or it could be game-changing breakthroughs in technology, perhaps the emergence of true artificial intelligence, or mastery over our genetic heritage. It could even be awe-inspiring images associated with our exploration of space, or visions of increasingly powerful spacecraft lifting off on missions to Mars and beyond.

Perhaps we're looking at a future where we need multiple equivalents of *Earthrise* to inspire us. Or maybe we need to be reinspired by that image of Earth rising above the horizon of the moon, and consider afresh the perils and possibilities in front of us.

As our space-facing technologies improve, there are plans afoot to send humans back to the moon and establish a permanent base there. As we do, we'll have another chance to look back on Earth from our celestial neighbor, and once again marvel at the beautiful, complex pale blue dot we live on. And as we look back on where we've come from, maybe we'll be inspired afresh to imagine what this precious Earth will be like fifty years from now, and how we can work together to achieve it.

From this future moon station, and maybe one day the red planet of Mars, we'll once again look out and see our future as a pale blue object, full of potential, hovering above the horizon of possibility. And as we contemplate this "futurerise," we'll renew our vows to generations to come, and the future they stand to inherit.

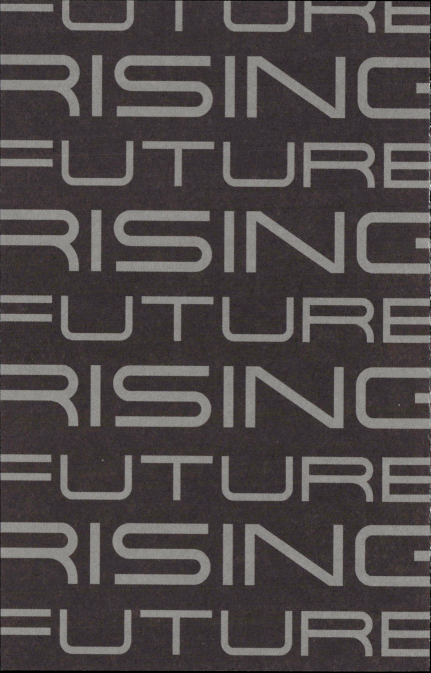

AFTERWORD

Despite our best efforts, the future will likely always remain intangible, ephemeral, constantly beyond our grasp. And yet, on this journey into what the future is and how we might think about it, I've found myself at times surprised, delighted, and humbled. There have been moments of serendipity as I realized just how deep the connections are between the way the universe works and how we feel about the future. And there's been a gnawing worry that our growing ability to mold and change the future continues to exceed our understanding of how to do this responsibly.

There's also been hope. When seen from the perspective of something that we can imagine, aspire to, and begin to design, the future reveals what exquisite and astonishing creatures we are. The fact that we can break away from the constraints of time's arrow and create a tomorrow that is different from today is quite astounding. Add to this our sense of justice and responsibility, our capacity to empathize with others, and our growing ability to map a course between what we know and what we imagine could be, and it becomes clear that we have the means to build a future that far exceeds the limitations of the present.

What hangs in the balance is whether we have the will to embrace this path forward.

The irony is that, while the universe conspired to create the conditions that led to the emergence of intelligent and visionary humans, it doesn't need us when all's said and done. We could be wiped out tomorrow, and there would still be a future—it just wouldn't be one that involved people. The universe doesn't care if we squander the talents we've been given—we're an aberration, a fortuitous but ultimately insignificant blip—in the grand scheme of things. But we *should* care. We've been given the gift of being the architects of our own future. And because of this, we have a responsibility to ourselves, and to future generations, to learn our trade and to build the best possible future we can imagine.

Whether we will collectively rise to this challenge remains to be seen. But for the future we have the potential to build together, I hope with all my heart that we do.

ACKNOWLEDGMENTS

Every book has an origin story, and this one began when my good friend and colleague Steven Beschloss asked if I'd consider writing a short book about the future as an everyday object. I may have laughed—I certainly thought it was an impossible task. Yet the idea stuck. And as I mulled it over, I began to see a way in which it might just work...

The result was this series of reflections—sixty in all, reflecting the passage of time between past and future. The initial scaffolding of thinking about the future as an object has all but gone—although you can still find the residual marks it left behind if you look closely. But it was a scaffolding that enabled me to transcend the original intent and create a much more expansive and personal narrative around how we think about the future, and our relationship to it.

Along the way, I received advice and feedback from a wonderful and dedicated group of colleagues, to which I'm deeply indebted. There was some tough love at times—as there should be—and I'm sure there'll be a reviewer or two who, on reading the final book, exclaims, "I cannot believe he wrote that!" Yet the advice I received was always helpful, and ultimately made the final journey a far more interesting and compelling one. Here, I need to especially acknowledge the help and support of Jason Brown, William Dabars,

Joshua Loughman, Nicole Mayberry, Becca Monteleone, JP Nelson, Randy Nesse, Martin Peres Comisso, Alycia de Mesa, Mateo Pimentel, Marissa Scragg, Natalie Severy, Bobby Sickler, Nikki Stevens, Steven Weiner, Jamey Winterton, and Dania Wright.

I also want to call out the work and support of everyone at Mango Publishing, and especially my wonderful editor Hugo Villabona, without whom this book wouldn't be what it is.

And of course, none of this would have been possible without the love, support, insights, and editing skills of my wonderful wife Clare—especially when asked for the nth time, usually late in the evening, "Could you just read this to see if it makes any sense?" Thank you!

ABOUT THE AUTHOR

Dr. Andrew Maynard is an author, former physicist, and leading expert on the socially responsible development of emerging and converging technologies. His PhD is in Physics from the University of Cambridge, UK. For over twenty years he has worked closely with experts from around the world on the challenges and opportunities presented by technologies ranging from nanotechnology and genetic engineering, to artificial intelligence and self-driving cars.

He was previously Chair of the World Economic Forum Global Agenda Council on Emerging Technologies, and continues to work closely with the Forum on beneficial and responsible technology innovation. In addition to his academic work, Andrew is a prolific writer, communicator, and sought-after speaker. He writes about technology and society on platforms ranging from Medium OneZero and Slate Future tense, to The Conversation, produces the YouTube channel *Risk Bites*, and is active on Twitter as @2020science. Dr. Maynard is currently a professor in the School for the Future of Innovation in Society at Arizona State University. He spends more time than any sane person should watching sci-fi movies. His first book is *Films from the Future*.

Mango Publishing, established in 2014, publishes an eclectic list of books by diverse authors—both new and established voices—on topics ranging from business, personal growth, women's empowerment, LGBTQ studies, health, and spirituality to history, popular culture, time management, decluttering, lifestyle, mental wellness, aging, and sustainable living. We were recently named 2019's #1 fastest growing independent publisher by Publishers Weekly. Our success is driven by our main goal, which is to publish high quality books that will entertain readers as well as make a positive difference in their lives.

Our readers are our most important resource; we value your input, suggestions, and ideas. We'd love to hear from you—after all, we are publishing books for you!

Please stay in touch with us and follow us at:

Facebook: Mango Publishing

Twitter: @MangoPublishing

Instagram: @MangoPublishing

LinkedIn: Mango Publishing

Pinterest: Mango Publishing

Sign up for our newsletter at www.mango.bz and receive a free book!

Join us on Mango's journey to reinvent publishing, one book at a time.